Stargazing

Stargazing

Peter Grego

ISBN-10: 0-00-719425-0
ISBN-13: 978-0-00-719425-0

ISBN-10: 0-06-081887-5 (in the United States)
ISBN-13: 978-0-06-081887-6

FIRST U.S. EDITION

Printed and bound by Printing Express Limited, Hong Kong

10 09 08 07 06 05
9 8 7 6 5 4 3 2 1

contents

introducing astronomy   6

eyes on the skies   24

recording the skies   42

our cosmic backyard   50

deep space   80

star charts & showpieces   100

Glossary   186

Need to know more?   188

Bibliography   189

Index   190

introducing

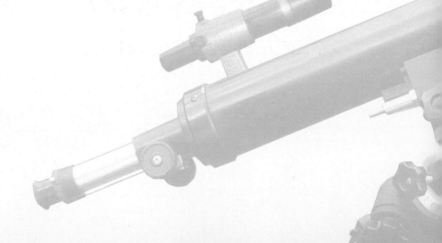

astronomy

Astronomy is the oldest science—written records of celestial events go back thousands of years. But people have gazed at the stars with wonder and awe ever since the first sparks of human consciousness flickered in the minds of our distant ancestors.

Heavens above

There is plenty to see in the night sky. Some things, like the constellations, are permanent fixtures; others, like the planets, move and change over time. A few phenomena, like aurorae, are spectacular but fleeting.

▲ Stargazers can forget about Earthly worries and hitch a ride on celestial photons.

Whether you use binoculars or a telescope, or even if you have no optical aid at all, there are enough sights in the night skies to keep the stargazer enthralled for a lifetime. This book, written for active stargazers eager to discover the universe with their own eyes, covers literally everything there is to be seen in the heavens, day and night.

Seeing the skies

So much can be seen in the skies without optical aid. Every stargazer ought to spend time learning the layout of the skies, the position of the main constellations, and the names of the brightest stars. This can only be achieved by actually standing under a starry sky and tracing the constellations with the aid of a star chart. The learning process cannot be completed in a single evening. During the course of a year, the heavens appear to revolve slowly around the Earth, and apart from those constellations near the celestial pole, their visibility is seasonal. Believe it or not, there are advantages to living in a light-polluted city. Since only the brighter stars can be seen, the skies appear less crowded, and the patterns of the main constellations are easy to trace. Under a dark rural sky, the heavens can appear so congested with stars that even experienced stargazers can become somewhat disorientated! Nothing stirs the soul more than the grandeur of the night skies seen from beneath a pristine dark sky.

Take a pair of sparkling eyes

Binoculars will reveal much more of the skies, and the impression of three dimensions in space—although wholly illusory—can be striking. The cratered surface of the Moon is revealed in all its glory through the smallest binoculars. Star colors are especially noticeable, and hundreds of double stars and deep sky objects, as well as countless glorious starfields, can be viewed.

Optic tubes and light buckets

With their larger light-gathering ability, telescopes will allow detailed, magnified views of the Moon and planets, as well as views of very faint objects like distant galaxies and nebulae.

Some transient phenomena, such as meteors and aurorae, are best enjoyed without any optical aid. Other phenomena, such as lunar eclipses, benefit from the low magnification afforded by binoculars. Certain phenomena require a telescope to be seen at all—for example, the sight of Jupiter's Great Red Spot transiting that planet's flattened cloud-streaked disk. The universe contains a panoply of glorious spectacles of varying magnitudes—whether it be the meteoric burnup of a dust grain 80 miles above our heads, or the sudden death of a star 80 million light years away. It's all there, waiting to be enjoyed.

▲ Comet Hale-Bopp dazzles in the skies above Stonehenge, Wiltshire, England—an ancient megalithic construction among whose purposes may have been to keep track of the year's celestial events.

▶ Brilliant Jupiter shines near the Beehive star cluster in Cancer.

A history of looking up

People have been systematically watching the skies and recording celestial events for at least 6,000 years. A lot has happened during that time, and human notions about the universe—influenced at first by religion, and in the last few centuries by science—have changed dramatically.

Cultures the world over have given rise to an amazing variety of cosmologies which granted the untouchable occupants of the heavens— the Sun and Moon, the stars and planets—a variety of supernatural powers. The gods of the sky were real personalities; each looked different, and each moved through the sky in its own unique way. Most powerful among the sky gods were the Sun and the Moon, sometimes interpreted as a competing pair of deities because of the phenomena of eclipses.

The movements of the heavens were believed to show the intentions of the gods. Mere mortals could learn the plans of the gods by carefully studying their movements, and in doing so recognize patterns which could be predicted in advance. By being able to predict celestial phenomena, humans believed that they had more control over events on the Earth. The remains of megalithic constructions like Stonehenge, whose stones are aligned with celestial points, are impressive evidence of how important it once was to be aware of sky phenomena. Astrology

▼ Eudoxus envisaged the Earth at the center of a series of transparent crystal spheres, upon which were fastened the Sun, Moon, individual planets, and stars.

INTRODUCING ASTRONOMY

arose from this potent mixture of careful observation and interpretation.

Greek astronomy

A more logical approach to understanding the universe arose in ancient Greece. Philosophers used their intellects to define the universe in physical terms. In the 4th century BC, Eudoxus devised a complete system of the universe. He envisaged the Earth at the center of a series of transparent crystal spheres, upon which were fastened the Sun, Moon, individual planets, and stars. Mathematics and geometry proved invaluable tools for understanding the scale of things. Using observation and trigonometry, Eratosthenes accurately measured the Earth's circumference (around 240 BC), and Hipparchus made the first accurate measurements of the Moon's distance and size (150 BC). Around the same time, Ptolemy compiled an encyclopedia of ancient Babylonian and Greek knowledge, including a definitive atlas of the stars—1,022 of them, contained within 48 constellations. Ptolemy expanded on Eudoxus' idea of the Earth-centered universe, explaining that the observed looping motions of the planets were produced by their making smaller circular movements as they orbited the Earth. These "epicycles" were incorporated into models of the Universe to preserve the notion that all celestial bodies possessed uniform circular motion. The epicyclic system increased in complexity, even featuring epicycles on epicycles to account for the observed movements of each planet.

▲ One of the major anomalies the Greeks had to solve is the Sun's altitude throughout the year which varies at midday from its highest at mid-summer to its lowest at mid-winter. The Earth's orbit around the Sun is elliptical, so the Sun's apparent speed (the time that it transits the local meridian) varies throughout the year, thus producing what appears as a figure-of-eight loop or an analemma.

MUST KNOW

Early interpretations
Certain events that could not be predicted—for example, the appearance of a bright comet, an aurora or a meteor storm—were, understandably, often regarded by our ancestors as signs of the gods' displeasure, portents of disaster, and impending doom for mankind.

▲ Uraniborg, Tycho's observatory on the island Hven, circa 1580. It was named after Urania, the Greek goddess of the sky.

Big Bang of reason

An explosion of scientific and astronomical inquiry began in 16th-century Europe. Ptolemy's theories, long held to be infallible by the powerful Church, began to be questioned, and many were found wanting. Nicolaus Copernicus (1473–1543) suggested that the Sun, not the Earth, lay at the center of the universe. This theory upset the Church which up to this point believed that humanity and all things human were at the center of the Universe. Thus, his literally revolutionary book promoting the heliocentric theory was not published until sometime after his death.

Tycho Brahe (1546–1601) was the last and greatest observer of the pre-telescopic era. His precise measurements of the stars and the movements of the planets, made with the aid of naked eye quadrants and cross-staffs, enabled his assistant Johannes Kepler (1571–1630) to place Copernicus' heliocentric theory on a firm scientific footing. Kepler formulated his three famous laws of planetary motion, based upon the fact that all the planets move around the Sun along elliptical paths.

An age of eye-opening discoveries

With the astronomical observations of Galileo Galilei (1564–1642)—made with the newly invented telescope—came astounding, undeniable proof of this radically different view of the universe's layout. Four small satellites were discovered orbiting Jupiter, and Venus showed phases, consistent with it being a globe in orbit around the Sun. The heavens were by no means perfect—the Sun often displayed irregular dark spots, and the Moon's surface was rough, cratered, and mountainous in places.

As telescopes improved, and a growing number of stargazers looked through them, it became clear that the Earth was just a planet with a satellite, orbiting the Sun between Venus and Mars.

Doubts were raised at the presumed special status of the Sun itself. Telescopes had already revealed that the Milky Way was composed of multitudes of stars that were too faint to be resolved individually with the unaided eye. If the stars themselves were like the Sun—but so far away that they appeared as mere points of light—then maybe the Sun did not lie at the hub of the cosmos, but was one of a broader mass of stars contained within the Milky Way. The abundance of new visual wonders prompted fresh ideas about the universe—ideas that seemed at odds with the traditional views of the Church.

▼ Galileo at the height of his powers in the early 17th century.

Great stargazers

As telescopes grew larger throughout the 17th, 18th and 19th centuries, familiar objects became better known and new objects loomed into view from the darker depths of the cosmos.

Johannes Hewelke (Hevelius) (1611–87) founded his own private observatory, equipped with many instruments that he designed and built himself. He made accurate measurements of star positions and produced the most advanced star atlas of its time, the *Uranographia*. Based upon his own observations, Hevelius also produced a detailed atlas of the Moon, the *Selenographia*.

Giovanni Cassini (1625–1712) was the first director of the Paris Observatory in 1671, the world's first fully equipped national observatory. He suggested that the mysterious rings of Saturn consist of a countless mass of tiny moonlets, and accurately calculated the distance between the Earth and Sun.

▲ The constellation of Taurus, from Hevelius' 17th-century star atlas.

In the late 18th century, comet-hunter Charles Messier (1730–1817) compiled a list of 110 deep sky objects that might, under various circumstances, be mistaken for new comets by other stargazers. Messier's list incorporates star clusters, nebulae, and galaxies, many of which appear as comet-like faint fuzzy patches through small telescopes. The list proved so useful that it is still referred to by stargazers.

INTRODUCING ASTRONOMY

14

The Herschels—a stargazing dynasty

▲ William Herschel, discoverer of Uranus, one of the greatest stargazers of all time.

Using telescopes that he had made himself, William Herschel (1738–1822) established himself as the world's most prolific stargazer. In 1781 he doubled the scale of the known Solar System when he discovered Uranus. Herschel went on to discover two more satellites around both Saturn and Uranus, in addition to recording hundreds of double stars and nebulae. One of Herschel's lesser-trumpeted, but in fact most significant discoveries, was the actual motion of the Sun and the Solar System through space—a finding which, along with his insights into the shape of the galaxy, demonstrated that the Sun was not at the center of the Universe.

Herschel built a telescope 40 feet (12 meters) long with a 49 inch (124 cm) mirror, mounted on a frame which revolved on a circular base; the instrument proved cumbersome to use, and he preferred a smaller instrument some 20 feet (6 metres) long. His sister, Caroline Herschel (1750–1848), was also a prolific observer. She discovered several nebulae and eight comets. John Herschel (1792–1871) expanded upon the work of his father. From South Africa he surveyed the southern skies, recording hundreds of new double stars and nebulae.

Julius Schmidt (1825–1884) was a 19th century visual observer who made a big impact on astronomy. A prolific observer, he drew the largest and most accurate map of the Moon ever based upon telescopic observations. The golden era of visual observation gradually came to an end with the arrival of astrophotography in the late 19th century, quickening the pace of discovery by enabling fainter objects to be recorded. The science of astrophysics developed as spectroscopes enabled the light from celestial objects to be split and analyzed, revealing the chemical constituents of the atmospheres of stars and planets.

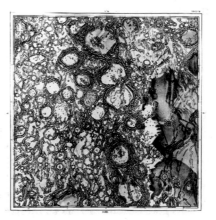

▶ The central portion of the Moon, mapped by Julius Schmidt in 1871, part of stargazing history's greatest lunar atlas.

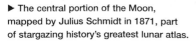

In infinite realms

How did the universe come into existence? What are energy and matter? What force guides the Moon in its orbit around the Earth, and the Sun around the center of the Milky Way? These are some of the fundamental questions that astronomers and cosmologists have attempted to answer.

▲ Albert Einstein had a profound influence on astronomy.

On the shoulders of giants

One of the greatest scientists in history, Isaac Newton (1642–1727), had an immense influence on our understanding of the universe. His invention of the Newtonian reflecting telescope helped us see the universe more clearly than ever before. Newton's invention of calculus gave astronomers a valuable mathematical tool with which to work out planetary orbits. Newton's three laws of motion appeared to underpin the mechanics of the cosmos. He also explained the force that guided the universe's movements in his law of universal gravitation. Newton realized that the force which compels an apple to descend to the ground, and the force which guides the Moon around the Earth, are one and the same thing.

▼ Around 14 billion years ago, the Big Bang created the universe. Matter was formed soon after, and within the first few billion years galaxies developed from clumps within this matter. The Sun and its planets were formed in the Milky Way galaxy around 5 billion years ago.

Einstein's universe

Newton's laws held sway for more than two centuries, until Albert Einstein (1879–1955) published his theories of relativity in the early 20th century. Einstein's famous equation $E=Mc^2$ (energy

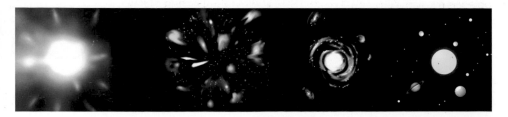

▶ Abell 1689, a cluster of galaxies whose gravity distorts the light from other galaxies far beyond it, seen here as a number of curving streaks.

= mass multiplied by the speed of light squared), showed that just a small amount of mass could be converted into a very large amount of energy—hence the enormous energy produced by nuclear reactions in stars. Einstein's special and general theories of relativity ventured into areas untouched by standard Newtonian physics. He maintained that gravity is the result of an object's mass curving the very fabric of space and time. Planetary movements around a really massive object like the Sun are governed by the extent of that space-time curvature, which lessens with distance. Planets near the Sun orbit very fast, while distant planets move around the Sun at a more leisurely pace.

Einstein predicted that light itself is bent by gravity—such effects have been seen in the form of gravitational lensing by galaxies producing distorted, and even multiple, images of other galaxies far beyond them.

An expanding universe

Spectroscopic measurements reveal that the light from distant galaxies is stretched out to the red end of the spectrum; such redshifts occur if an object is moving away from the observer. Edwin Hubble (1889–1953) demonstrated that the further a galaxy is, the greater its redshift, which implies that the Universe is expanding. Cosmologists conclude there must have been a time in the distant past when all the matter in the universe was bunched up together at a single point. According to the Big Bang theory, the starting point containing all the energy and matter of the Universe exploded around 14 billion years ago.

▶ Spaceship Earth

Our home planet, a globe some 7,945 miles in equatorial diameter, is the third big rock from the Sun. From our spinning sphere, the skies appear to parade around us every day.

MUST KNOW

Northern hemisphere solstices

The Sun reaches winter solstice around December 21, marking the start of winter, and the summer solstice around June 21, heralding summer.

Time and season

Rotating on its axis once in 24 hours, the Earth orbits the Sun once every 365.25 days—the extra quarter day is saved up and added on every four years in February, giving us an extra day, known as a leap year.

The Earth's axis is tilted at an angle of 23.5° to the plane of its orbit around the Sun (the ecliptic—see below), and this tilt is constant with respect to the stars. This gives us the phenomenon of the seasons, whose effects are more noticeable the further away from the equator you go. In December, the Earth's north pole is angled away from the Sun, and during the northern winter the Arctic regions are in permanent darkness. December in the opposite hemisphere brings the height of southern

▼ Illustrating how the Earth's seasons are caused by the tilt of the Earth's axis to the plane of its orbit around the Sun.

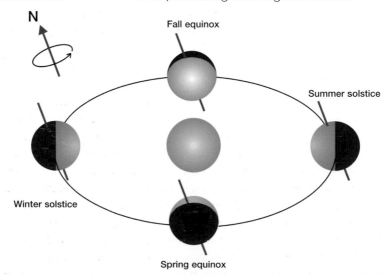

Fall equinox

Summer solstice

Winter solstice

Spring equinox

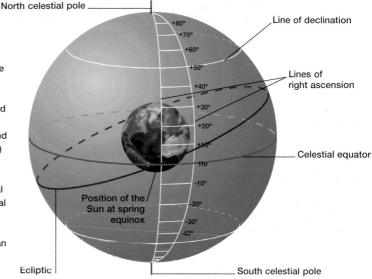

North celestial pole

Line of declination

+80°
+70°
+60°
+50°
+40°
+30°
+20°
+10°

Lines of
right ascension

1hr

Celestial equator

-10°
-20°
-30°
-40°

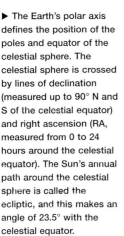

Position of the
Sun at spring
equinox

Ecliptic

South celestial pole

▶ The Earth's polar axis defines the position of the poles and equator of the celestial sphere. The celestial sphere is crossed by lines of declination (measured up to 90° N and S of the celestial equator) and right ascension (RA, measured from 0 to 24 hours around the celestial equator). The Sun's annual path around the celestial sphere is called the ecliptic, and this makes an angle of 23.5° with the celestial equator.

summer, when the Antarctic regions are bathed in 24-hour sunlight. It is still very cold there, though, no matter what time of the year it is! Precisely the opposite happens six months later when, during June, the Earth has moved to the other side of its orbit around the Sun. The north pole is angled towards the Sun, producing the height of northern summer, while it is mid-winter in the southern hemisphere. These two extremes are called the winter and summer solstices.

Equidistant between the solstices are the spring and autumn equinoxes, when neither pole is angled towards the Sun, and all parts of the Earth experience 12 hours of sunlight and 12 hours of darkness.

The ecliptic

From our perspective, the Sun appears to move steadily through the heavens day after day, proceeding along a well-defined path called the ecliptic which passes through all of the 12 original constellations of the zodiac. If the Earth's axis was perfectly upright with respect to the ecliptic, the celestial equator and the ecliptic would coincide; all days would have 12 hours of sunshine, and the Sun would rise due east, and set due west, each and every day, no matter where you lived on the Earth. But because the Earth's axis is tilted to the ecliptic, the Sun's rising and setting times and directions, and its height above the southern horizon at midday, vary through the year, with extremes at the two solstices. The Moon and seven of the major planets also stick quite closely to the ecliptic, although

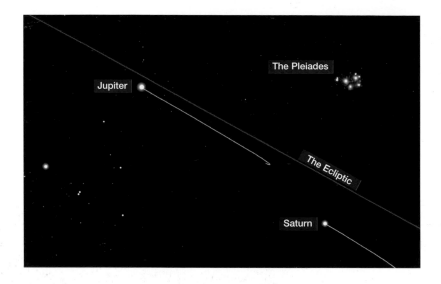

▲ The paths of Jupiter and Saturn through Taurus during early 2001.

their orbital planes display a varying amount of tilt which takes them several degrees away from it at times.

The stars' annual parade

From any particular part of the Earth there is a region surrounding the celestial pole which is always visible, and the extent of this circumpolar region varies with latitude. From the north pole, exactly half the sky—everything above the horizon—is circumpolar. From the latitude of, say, London at 52° N, everything north of 38° N on the celestial sphere is deemed to be circumpolar.

Viewing the other constellations depends largely on the season. As the Sun proceeds along the ecliptic, so formerly invisible constellations begin to emerge into the eastern morning skies, reaching their highest above the horizon at midnight around six months later.

MUST KNOW

During the Sun's annual circuit along the ecliptic, its glare renders the constellations immediately surrounding it practically impossible to view.

Constellations then sink gradually into the evening skies, moving ever westward until they are once again lost in the evening afterglow and the glare of the Sun.

Find the Pole Star

Polaris, the Pole Star, is located within just 1° of the north celestial pole, and it appears more or less stationary throughout each night. It is not a particularly bright star, but it can usually be seen

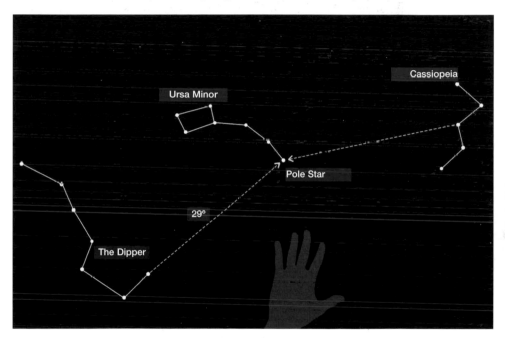

▼ Polaris, a bright star close to the north celestial pole, can be found using the two "pointer" stars of the Dipper or by using the W of Cassiopeia.

from even the most light-polluted of sites. Polaris is useful to locate in order to orientate yourself. Dubhe and Merak, the two end stars in the Big Dipper (the seven brightest stars in Ursa Major), point towards Polaris. If these are too low to be seen, an alternative is to use the W shape of Cassiopeia (on the opposite side of the pole to the pointers); the middle stars of the W form a broad arrow pointing in the general direction of the pole.

Enjoy the skies

Stargazing is a truly out-of-this-world hobby. Nothing on Earth compares with contemplating the night skies. Whether you're a beginning stargazer, or a seasoned sky watcher, the skies provide an endless source of fascination.

The joy of stargazing

Getting to know the brighter stars and constellations is an essential first step along the learning curve to becoming a well-rounded stargazer. The charts in this book (see pages 100–185), made to a generous scale, clearly show all the naked eye stars visible under dark skies, as well as hundreds of deep sky objects which can be viewed through binoculars and telescopes.

Once you understand the basics of observing, and the types of celestial objects and phenomena visible, you will soon discover the limitations of the instrument you use, and the limitations imposed by your particular observing site. The good news is that no matter where you live, and regardless of the equipment available to you, the skies have plenty to offer. New stargazers are struck by just how much there is to be seen, even from urban locations.

Images of space taken by big observatories can be colorful and spectacular. Stargazers usually satisfy themselves with much less color and detail in the objects they view. What often thrills and enthralls the stargazer is a combination of the sight of an object or phenomenon (however dim and featureless it may seem to the uninitiated), with a scientific knowledge of what it really is. A faint star in the Andromeda Galaxy may not appear to deserve a second glance—but when it is realized that

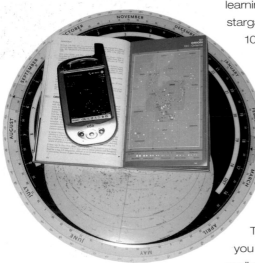

▲ You can plan your stargazing with a planisphere, star map, or computer program. PDAs like the one shown are excellent for consulting outdoors, since they are backlit, and easily read in the dark.

the star is a supernova, the light produced by an exploding star—then the star becomes worthy of more prolonged scrutiny.

Stargazing Zen

Many stargazers have a deep reverence for their equipment—whether it's a budget pair of 7x30 binoculars or an advanced computerized telescope. Some stargazers even have pet names for their telescopes! Through the lenses of the magic tube, streams of photons from distant stars and planets are focused onto the stargazer's keen retina. Observers may become secret captains of their own starships, regarding the eyepiece as a porthole or the viewscreen on the bridge of their spacecraft. At a time of night when others sleep, stargazers tour the cosmos. You know when you are in love with stargazing when you begin to feel guilty for not being out there whenever there is a clear night!

▲ When sunset comes, stargazers eagerly set up their telescopes and prepare for an evening of cosmic exploration.

Stargazing tips

The most enjoyable stargazing is done when the observer feels comfortable and safe. If you are hunting constellations and deep sky objects, it's best to have some idea about your intended quarry by consulting a planisphere or star chart beforehand. Wrap up well with fresh dry clothes —even on warm summer nights, the elements can slowly chill you. Be familiar with your observing site—even your own backyard can seem like an alien world in the dark, full of unsuspected dangers. An adrenalin rush is not uncommon, and it is normal to tremble and become more aware of background noises— that is just a survival instinct that we have inherited from the time our distant ancestors were frightened mammals. Take your time to get everything set up—after all, stargazing is supposed to be an enjoyable hobby!

want to know more?

Take it to the next level...

Go to...
▶ **Dark adaptation**—page 26
▶ **Getting your bearings**—page 40
▶ **Recording observations**—pages 42–49

Other sources
▶ **Planispheres**
   find out what's up and when
▶ **Astro magazines**
   keep up to date with sky events
▶ **Days out**
   get inspired by space places
▶ **Join a local astro club**
   meet other enthusiastic stargazers
▶ **Astronomy holidays**
   relax under dark country skies

eyes on

the skies

You don't need expensive equipment to enjoy stargazing. You can identify constellations, locate planets, and follow the Moon's phases with the unaided eye alone. Binoculars reveal more of the night skies, and a telescope will give you a close-up view.

▶ The eyes have it

Our eyes are our interface with the universe, and if you know a few things about how they work, then you can make the most out of viewing the heavens.

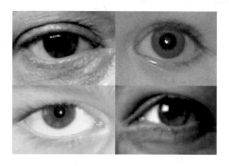

▲ Eye color has nothing to do with the quality of your eyesight, but vision becomes less acute with age.

The human eye

The eye has a much more complex structure than any camera. Light is focused by the cornea, a membrane at the front of the eye, into the pupil, and through the lens behind it. Muscles alter the shape of the lens, allowing it to focus the light onto the retina inside the back of the eyeball. Light-sensitive cells in the retina convert light into electric signals, which are sent to the brain and processed into an image.

Dark adaptation

In the dark, pupils dilate to their maximum size, allowing the maximum amount of light into the eye. Stargazers can take advantage of dark adaptation, but it takes time for pupils to dilate. Step outside from a bright room into a dark backyard, and it may be difficult to see any stars at first. After a while, stars begin to be seen, and after about half an hour in darkness you are able to see stars to the limit of your vision. But be warned! Any bright light—be it automobile headlights or a neighbor's "security" light—will instantly ruin your dark adaptation.

▼ The eye has two gel-filled chambers and two main optical components, the cornea and lens.

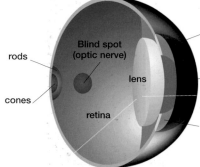

rods

Blind spot (optic nerve)

cones

lens

retina

iris

pupil

cornea

Light pollution

The night sky can be an awesome sight. From a dark location on a moonless night, around 1,500 stars can be seen with the unaided eye at any one time. At times the Milky Way—millions of stars in our own galaxy, so far away that they merge into a mottled band—

can be glaringly obvious. Faint nebulous patches can be spotted here and there. Some of those are star clusters, others are vast clouds of glowing gas, while a few are galaxies far outside the Milky Way. But if you live in a big city, the night skies will never appear truly dark. Stray light from homes, businesses, industry, and streetlights illuminates dust and moisture in the air, causing skyglow. This drowns out faint stars and low-contrast objects like the Milky Way, but there are still plenty of objects that can be viewed to their full, such as the Sun, Moon, and planets.

▲ The Moon, a few bright stars, and planets may be the only celestial objects visible in the skies above a light-polluted city.

Avert your gaze

There are two types of light-sensitive cells in the retina. Cone cells at the center of the retina give us detailed color vision at the center of our view, but they are only triggered by bright light. In the dark, only the rod cells lying around the cones, away from the center of the field of view, are triggered. That is why it is hard to see a dim object when you look straight at it—you need to look a little way to the side of its actual position—a useful technique that stargazers call "averted vision." The rods do not deliver detailed images, and they cannot distinguish colors—which is why a landscape illuminated by the Moon appears somewhat lacking in color.

▼ This photograph of the constellation of Cygnus approximates the naked eye view from a dark site.

▶ Binoculars

With their portability, wide fields of view, and low magnification, binoculars give the stargazer unlimited freedom to roam the skies. By using both eyes to view the skies, we get a feeling of really "being there" among the stars.

▲ A pair of 7x50 binoculars can be comfortably hand-held for general stargazing.

▲ A tripod can be used for stargazing with big binoculars, but viewing objects high in the sky may require some craning of the neck!

Steady goes

Binoculars range in size from tiny palm-sized glasses to big beasts that can only be used effectively if they are mounted on a sturdy tripod or special binocular mount.

Most binoculars are light enough to be hand-held without difficulty, but to enjoy the view to the full, it's best to hold binoculars as still as possible while maintaining a comfortable posture, minimizing binocular wobble, and image shake. The stiller the view, the greater number of faint objects can be seen. Many stargazers find that being seated, or even lying down in a garden recliner, greatly enhances their enjoyment of binocular viewing.

Image stabilized binoculars use inbuilt electronic sensors to keep the view shake-free. They are expensive compared with even premium quality "regular" binoculars, but many stargazers find them great to use because they don't require a binocular support.

Binocular types

Simple binoculars like opera glasses (sometimes called "Galilean" binoculars) are not really suitable for astronomy, as they have a small field of view and the image they deliver is distorted in a number of ways. Most good binoculars use internal glass prisms to fold the light, making them shorter and more compact. Classic Porro prism binoculars are W-shaped, with two widely spaced objective lenses. Roof

prism binoculars are generally more compact and lightweight, and resemble two small telescopes side by side.

Size and power

The power of binoculars is identified by two figures that tell us their magnification and the aperture of their objective lenses. 7x50s magnify 7x and have 50mm objectives. For general stargazing, 7x50s are ideal. They are reasonably lightweight, deliver a wide field of view, gulp in lots of light, and have a magnification low enough not to make any binocular wobble obtrusive when hand-held for short periods. 7x50s give a field of view about 7° across (14 times the diameter of the Moon).

Choosing binoculars

If bought from a reputable optical dealer, your binoculars are likely to be optically sound. Usually you get what you pay for, but that does not mean that the least expensive binoculars will not deliver great views of the night sky. Beginners are advised to start with lightweight, low-magnification binoculars such as 7x50s. Avoid heavy, high magnification, or variable-magnification binoculars which exceed your needs as a novice stargazer.

From front to rear: 25mm, 50mm, and 70mm binoculars.

▶ Refractors

Asked to imagine a telescope, most people think of a refractor—a long tube with a lens at the front, and an eyepiece at the other end. Refractors are versatile, robust, and need little maintenance, compared with other telescope types.

Bending light

Refractors use an objective lens to bend (refract) the incoming light to a focal point. The eyepiece magnifies the image, and the distance of the eyepiece from the objective lens can be adjusted to deliver a focused image to the eye.

Galileo magnifico

In 1609, Galileo made wonderful discoveries when he turned a home-made refractor on the heavens. One of Galileo's refractors had an objective lens of just 30mm and a single eyepiece lens that magnified x20, but it was enough for him to discover a host of previously unseen (and unsuspected) objects. Today, Galileo's design is considered rather crude, because the images it produces contain lots of false color, known as "chromatic aberration." Many low-quality, low priced telescopes are based on the Galilean design and should be avoided. Galilean telescopes produce images with colored fringes (most noticeable around bright stars and the Moon), because a single

▲ Galileo's telescope revealed the Moon's craters and Jupiter's four main satellites.

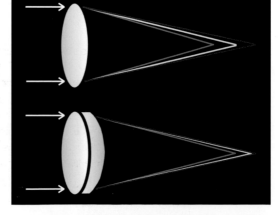

▶ A single lens (top) does not focus all the colors in light to a single point. An achromatic lens does a better job overall in focusing the light.

lens cannot focus all the light to a common point. White light is made up of a rainbow of colors, from red light through to blue light—from long to short wave.

Astronomical refractors

Most astronomical refractors overcome the problem of annoying amounts of false color by using an objective lens made up of two specially shaped pieces of glass which bend the light that passes through them into as near a common focus as possible. These "achromatic" refractors do not get rid of all the false colors, however. "Apochromatic" refractors have three or more lenses, sometimes made of exotic glass like fluorite. Apochromats are very expensive, but they virtually eliminate false color, so are worth investing in if you are serious about stargazing.

▲ A special filter covering the end of this refractor allows the Sun to be viewed safely (see pages 56–7 for viewing advice).

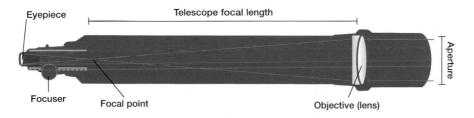

▲ Cross section through an achromatic refractor.

MUST KNOW

Gathering light

The larger a telescope's aperture (the size of its objective lens or primary mirror), the more light collects. Deep space objects, like nebulae and galaxies, are very faint, so light gathering ability is important to anyone interested in viewing these. The bigger the aperture, the brighter the object appears through the eyepiece. A 200mm telescope has four times the light gathering ability of a 100mm telescope. A 100mm telescope will reveal stars down to around magnitude 12 (1,000 times dimmer than the faintest star visible with the unaided eye), while a 200mm telescope will reveal stars around five times fainter (magnitude 13.5).

► Reflectors

As their name suggests, reflecting telescopes do most of their work with mirrors. There are a number of different reflector designs, but the one most loved by stargazers is the "Newtonian" reflector.

On reflection

Isaac Newton invented the reflecting telescope in 1671. Newtonian reflectors use a large mirror at the bottom of the tube to collect and focus light, while a small flat mirror near the top of the tube reflects this light out of the side of the tube into an eyepiece. Newton's design got rid of the false colors that had plagued refractors. Early reflectors used mirrors made out of polished speculum (a metal alloy), but these tarnished quickly. Speculum mirrors were superseded by silvered glass mirrors, but today aluminized mirrors are used, some with coatings that keep the surface from tarnishing.

▲ Newton's handcrafted reflector, presented to the Royal Society.

Cassegrains

Another type of reflector, devised by the French sculptor Guillaume Cassegrain just after the Newtonian, uses its secondary mirror to reflect the light back through a hole in the primary into an eyepiece behind it. Cassegrain reflectors are

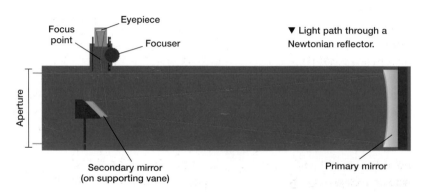

Focus point

Eyepiece

Focuser

Aperture

▼ Light path through a Newtonian reflector.

Secondary mirror (on supporting vane)

Primary mirror

Maksutovs

A different type of cat, using a deeply curved correcting lens rather than the relatively flat correcting plate of the SCT, was designed in 1941 by Dimitri Maksutov. The curved lens, called a "meniscus," gets rid of a type of distortion in the focus of light known as "spherical aberration," producing a flatter image. Instead of having a secondary mirror attached to a cell mounted in the correcting lens, like SCTs, most Maksutovs have an aluminized spot central to the inside surface of the meniscus, which acts as the secondary. Maksutovs are renowned for their high contrast, high resolution views that can equal the performance of similar sized apochromatic refractors.

▲ Lightweight and powerful, this 127mm Maksutov is an ideal travel 'scope.

MUST KNOW

Cool it

It's tempting to use a telescope from the warm indoors, the tube pointed though an open window—but the image is not likely to be good. Warm air escaping from the room produces turbulence, making the image shimmer uncontrollably. When taken outdoors, telescopes (especially cats) require at least half an hour of cooling time in order for them to perform well.

▼ Manual focusing with both SCT and Maksutov telescopes usually entails turning a small knob at the back of the instrument.

More cats

Variations on both the Schmidt and Maksutov designs have been combined with the Newtonian to produce the Schmidt-Newtonian and Maksutov-Newtonian telescope. The secondary mirrors in both divert light out of the side of the tube into the eyepiece rather than through a hole in the primary. All cats are closed systems, and if they are treated well, their mirrors can remain in excellent condition for decades. The front lens of a cat is prone to dewing after a while under the stars. Dew can be kept at bay with a dew cap or a commercial electrical anti-dewing strip.

▶ Eyepieces & accessories

Having a good telescope is only half of the story. To get the most out of stargazing, you need several good quality eyepieces and one or two items that enhance your enjoyment of observing the skies.

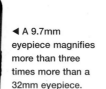

◀ A 9.7mm eyepiece magnifies more than three times more than a 32mm eyepiece.

Eyepieces

Once light has been collected and focused by the telescope's objective lens or primary mirror, the job of the eyepiece is to magnify the image and deliver it to the stargazer's eager eye. Most new telescopes bought from reputable optical dealers are supplied with a good quality medium power eyepiece or several eyepieces giving a range of magnifications from low to high. But there is little point in having an optically superb telescope and using a low quality eyepiece at the other end.

Eyepieces come in three barrel diameters— 1in, 1.25in, and 2in. Most telescopes take 1.25in eyepieces, which are by far the most widely used type. The 2in eyepieces are big and heavy, reserved for more expensive low-power, wide field designs. A 1.25 to 2in eyepiece adaptor is often required to use these big eyepieces in most telescopes. Budget instruments may only accept 1in eyepieces of very basic design. Basic eyepieces include Huygenian, Ramsden, and Kellner types (all centuries old designs), but they're rarely labelled as such—more likely they are touted as "modified" or "super achromat" eyepieces. Often mounted in plastic tubes, they all have tiny fields of view, and are frustrating, to use— like looking down a long tunnel. They ought to be avoided.

MUST KNOW

Magnification
To calculate the magnification given by any particular eyepiece, simply divide the telescope's focal length by the focal length of the eyepiece. A 1,000mm focal length telescope gives a magnification of x100 when used with a 10mm eyepiece.

Plössl eyepieces are the most popular type of eyepiece, and most stargazers have one or more of them in their kit. They range from very short focal lengths of 4mm (high magnification), to long ones up to 40mm (low magnification). Their apparent field of view is a generous 50°, with some more expensive varieties having wider fields. Longer focal length Plössls have better eye relief, and are more comfortable to use. Spectacle wearers require good eye relief eyepieces if they wish to wear glasses while viewing. With huge apparent fields of view (70 to 80° +), ultrawide eyepieces deliver a so-called "spacewalk" experience, where the user is oblivious to the edge of the field when looking at an object near its center. Budget ultrawides include the Erfle eyepiece, while Nagler eyepieces are expensive, but optically superb.

▲ Zoom eyepieces deliver a range of magnifications at the twist of a barrel—ideal if you are on a budget or are tired of carrying a whole box load of different eyepieces to the telescope.

Star diagonals

Sometimes the eyepiece gets into an awkward position to observe from. That is where a star diagonal comes in useful. Inside the diagonal, a small mirror or prism diverts the light at right angles, enabling more comfortable viewing. The image is reversed, but that's a small price to pay.

Barlows and reducers

Your range of eyepieces can be virtually doubled by using a Barlow lens, a useful accessory that lengthens the telescope's effective focal length, boosting the magnification delivered by any eyepiece; 2x and 3x Barlows are the most common types. An advantage of using a Barlow is that it improves eye relief in Plössl eyepieces, making high-powered viewing better. Focal reducers are the opposite of Barlows. As their name suggests, they reduce the telescope's effective focal length.

▼ Many stargazers consider the Barlow lens an indispensable tool.

▶ Telescope mounts

To get the best out of any telescope, its mount needs to move freely and smoothly, and it requires a stable platform, free from wobble and shake.

▲ Dobsonians are so simple that many stargazers choose to build their own, like this superb example of an open tube 200mm Newtonian.

Altazimuth mounts

Altazimuth mounts allow the telescope to move up and down and side to side at the touch of a finger, twiddle of a knob, or push of a keypad button. An altazimuth-mounted telescope needs adjusting regularly in both axes to keep astronomical objects in the field of view, as the Earth's rotation makes them appear to move from east to west across the sky.

Simple altazimuth mounts are often provided with small refractors. They can be awkward to use, especially at high powers when objects appear to drift across the field of view at a lightning pace, so it is best to stick to low magnifications when using them. Some altazimuth mounts have handy slow motion controls that allow objects to be tracked without touching the telescope tube.

Dobsonian mounts

These are the easiest to use. They have big bearings that ride on teflon, a silky smooth plastic that allows fingertip control. Dobsonians are versatile and they can accommodate reflecting telescopes up to a hefty 500mm diameter or larger, while retaining their ease of use—although stepladders may be required to get to the eyepiece of a really big Dobsonian!

A number of altazimuth telescopes are provided with electronic keypad control, allowing the slewing speed to be varied. These are fun to use and allow the stargazer to enjoy viewing the object rather than worry about pushing the telescope around manually.

Equatorial mounts

Once the polar axis of an equatorial mount is accurately aligned with the pole, a celestial object centered in the field of view can be kept there by moving the telescope from east to west in pace with the apparent movement of the sky.

German equatorial mounts

GEMs can be heavy, as they need counterweights to balance the telescope. Most GEMs can be fitted with an electric drive on the polar axis to track celestial objects. A hand controller and an extra drive can be fitted on the declination (up-down) axis of most GEMs, allowing the telescope to be slewed in any direction at the push of a button.

Fork mounts

These are provided with most SCTs; indeed, most SCTs are integral to the mount and cannot be removed from it. Slung between the forks of the mount, the short-tubed SCT is comfortable to use, since the eyepiece ends up in fewer awkward positions than a GEM-mounted telescope.

Tripods

Stargazers who store their telescopes indoors or in a backyard shed normally have a sturdy tripod on which to place their telescope. One drawback with the portable approach to observing is having to set up from scratch before each observing session. For general stargazing, an equatorial mount can be roughly aligned with the celestial pole using the pole star as a guide—having the polar axis within a couple of degrees will satisfy most observers.

▼ A 150mm refractor on a German equatorial mount atop a permanent pier.

▶ Location, location, location

The sky is a big place, and our view through binoculars and telescopes is awfully small in comparison. While the Moon and bright planets may be easy to locate, it is much more of a challenge to hunt for faint celestial objects.

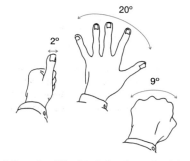

▲ It can be difficult to judge the scale of the sky by just looking at a map, so use your hand to approximate sizes. All around the celestial equator measures 360°.

First, get your bearings

The entire night sky is called the celestial sphere, and it is divided into 88 constellations. Some span a wide swathe of sky, while others are small enough to be easily obscured by the outstretched hand. Most constellations visible from the northern hemisphere were designated in ancient times and have Latin names, like Ursa Major (the "Great Bear"), while many of the bright stars have names that derive from Arabic sources, like Algol (the Demon Star). Once the names and locations of the brighter stars and constellations are known, they can be used as pointers to less conspicuous celestial objects.

▼ The "Big Dipper" of Ursa Major is a prominent pointer to other celestial objects in its vicinity.

RA and Dec

Just as we locate our position on the Earth with latitude and longitude, astronomers use a similar system of co-ordinates to pinpoint celestial objects. Right Ascension (RA) equals longitude on the celestial sphere, while Declination (Dec) equals latitude. RA is divided into 24 hours, each hour divided into minutes (') and seconds of arc (").
Dec is measured from the celestial equator to the pole (0 to 90°), and has a negative value for objects to its south. Star maps are usually marked with a scale of RA and Dec. Knowing how to locate an object based on its co-ordinates is extremely useful.

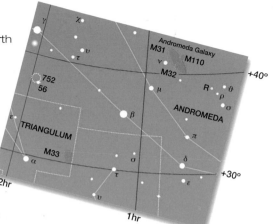

▲ The region of Andromeda and Triangulum hosts the two nearest large spiral galaxies, M31 and M33. M33 lies at 01h 34m RA, +30° 39' Dec. Can you spot the comet at 02h 03m RA, +41° 00'?

Star hopping

Bright stars near a fainter celestial object can be used to navigate to it, a technique called "star hopping."

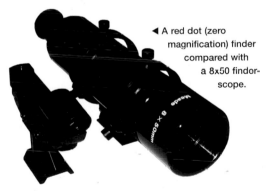

◀ A red dot (zero magnification) finder compared with a 8x50 findorscope.

It is important to have your telescope's finderscope accurately aligned with the main instrument. A red-dot finder will give an unmagnified right way up view, but you will need to turn your star map upside down to match the view through a conventional finderscope. Once an object's position is centered in the finder, use a low-power eyepiece to zero-in on it.

want to know more?

Take it to the next level...

Go to...
▶ **Constellations**—page 102
▶ **Viewing limits**—pages 104–5
▶ **Star charts**—pages 108–85

Other sources
▶ **Star atlases**
a closer look at the skies
▶ **Computer programs**
discover the heavens in detail
▶ **Internet sources**
learn more about celestial navigation
▶ **Attend a star party**
enthusiastic stargazers will be your guide
▶ **Set easy targets at first**
search for the brightest objects initially

recording

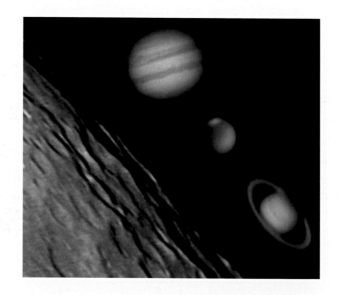

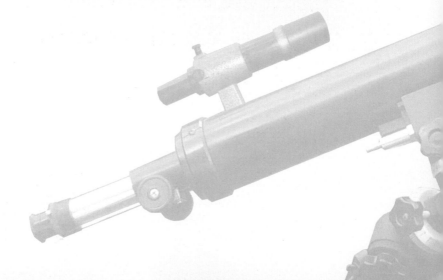

the skies

Just seeing the heavens is satisfying enough, so the idea of sketching while observing in the dark might seem like a lesson taught by a weird nocturnal art school. But ever since Galileo put pencil to paper, stargazers have wanted to record their views of the night skies.

Drawing on experience

Making notes and sketches of astronomical objects can enhance your enjoyment of astronomy. Practiced regularly, it improves skills of perception, as the observer is learning to attend closely to the fine detail visible through the eyepiece.

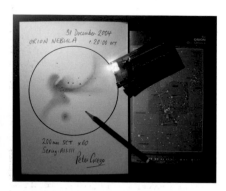

▲ Armed with a good star map, a dim red torch, pencil, and paper, the stargazer can make a permanent record of his or her telescopic journeys through space.

You don't need to be a good artist to be able to make satisfying observational drawings. Indeed, the experience of drawing at the telescope— sitting in the dark, using one eye to squint through an eyepiece, and then using both eyes to sketch upon a dimly-illuminated piece of paper —would challenge most skilled artists unused to the technique. Experience counts above all, and this is gained through regular practice.

Materials

A sketchpad of smooth thick cartridge paper, say A5 size, is easy to handle at the eyepiece. A soft dark pencil lower than grade 2B is recommended, since these are easier to apply and erase than harder grades. A harder pencil may prove troublesome to apply to slightly damp paper, which often happens as the dew settles at night.

A small hand-held torch with a dim red light should be sufficient to illuminate the sketchpad, especially if you are attempting to draw faint nebulae and galaxies. Low-intensity red light does not adversely affect dark adaptation, while the bright white light of a regular torch will make your pupils contract so much that the object you are observing might no longer be visible through the eyepiece.

Deep sky smudges

When observing deep sky objects, pre-draw large circles on the paper to correspond with the eyepiece's field of view—a CD makes a great

MUST KNOW

Captain's log

Note the date and time of your observation, in addition to the instrument and magnification used, and the seeing conditions. It would also be useful to make notes of any interesting or unusual features that might not be so obvious in your sketch.

template to draw around. Nebulae and galaxies appear as faint patches through the eyepiece. Structure within the object will be easier to see using averted vision. Once your deep sky quarry is centered in the field of view, plot the brighter stars, keeping them reasonably small and round. When drawing a nebulous object, use light strokes to mark its outline and loosely apply layers of pencil. Remember that you are sketching in negative—brighter areas will be darker on your sketch. With the tip of your finger, smudge the drawing so that the pencil strokes are blended in. Lastly, dot in any really faint stars. Indoors, an eraser can be used to emphasize any dark areas visible within your object, such as dark lanes within a galaxy. Comets can be depicted using much the same technique.

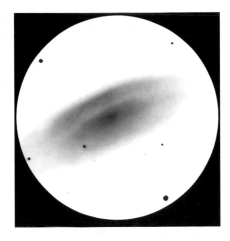

▲ Observational drawing of the Andromeda Galaxy.

Planetary blanks

Planets seldom appear perfectly round. Mercury, Venus, and Mars nearly always display a phase, and drawings of them are usually made on predrawn blanks of 2in. diameter. Jupiter is a distinct elliptical shape and Saturn has a complex set of rings surrounding it, in addition to being elliptical itself. Printed blanks for drawing Jupiter and Saturn can be obtained from a number of astronomical society observing sections, such as ALPO in the USA and the SPA in the UK. Blanks can also be downloaded from the Internet, or even prepared and printed using a suitable PC program.

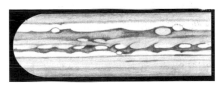

▲ Jupiter's rapid rotation means that features in its atmosphere rapidly transit across the planet's imaginary central meridian. This strip drawing, made during a five hour-long observing session on December 14–15, 1990, saved preparing four or five whole disk drawings.

▶ Observational drawing of the lunar crater Bullialdus.

Sketching the Moon

So much fine lunar detail is visible in the smallest telescope that it might appear futile to consider making a drawing. By concentrating on depicting a small area—say, an individual crater or small group of features—the observer will find that the task becomes defined and manageable.

Celestial snaps

For more than a century and a half, astronomers have refined their photographic techniques to capture ever more detailed images of the farthest reaches of the universe. But you do not need access to an advanced space telescope to take superb photographs of the night skies. Basic photographic equipment can produce some really stunning celestial images.

A dark night sky can appear ablaze with dozens of bright stars, with hundreds of fainter stars down to the limit of visibility and the glowing tract of the Milky Way dividing the sky in two. For all its visual magnificence, a basic "point and shoot" compact camera is not very effective at capturing what can be so plainly seen.

Regular photographic film is hundreds of times less effective at registering dim starlight than the human eye. When exposed to light, film undergoes a chemical alteration, as microscopic silver-halide crystals within it are prompted to grow. An image's intensity depends on the sheer amount of light impacting the film, and standard film is not really made to record the low light levels of night time scenes. This problem can be overcome by using more sensitive film and exposing it for longer periods. Regular film for everyday use is usually rated 200 ISO. A higher ISO number means a greater sensitivity to light, and most camera stores sell anything up to 1,000 ISO.

▲ Undriven photograph of Comet Hale-Bopp in April 1997. Including a natural foreground adds interest to an astronomical image.

Unguided photography

Most standard 35mm film cameras have a facility to take long-exposure images. Set to infinity, mounted on a tripod

◄ Undriven photograph of Orion, taken with a standard 35mm compact camera.

and aimed at a dark starry sky, an exposure of five minutes should be sufficient to record stars down to naked eye visibility. Earth's rotation causes each star image to appear slightly trailed. An exposure aimed at Polaris, near the north celestial pole, will show that each star trails an arc centered around the pole, and the further away from the pole, the longer each trail appears.

Unguided images are ideal for capturing aurorae, meteor trails, and trails left by artificial satellites. Aurorae can be bright enough to photograph from urban locations, but their constantly changing forms may require short exposures. To capture meteors, a camera aimed toward the shower's radiant on the night of their predicted maximum ought to record one or two of the brighter meteors, and perhaps—if you are lucky—a really bright meteor called a fireball. Artificial satellites orbit along predictable tracks, and a few of them, like the International Space Station, can appear dazzlingly bright along part of their track. Predictions for bright satellites can be found on various sources, including www.heavens-above.com.

Exposures are limited by light pollution, and an exposure of 15 minutes may be all that can be achieved from an urban area without the skyglow of artificial lights casting an undesirable hue over the resulting image.

Experiment
Photographing the skies successfully requires a degree of trial and error, so be prepared to go through a few rolls of film during the learning process. Astrophotographers usually bracket their images, taking a range of exposures during each imaging session rather than hoping that the one exposure will capture the scene satisfactorily.

▼ A 127mm Maksutov telescope set up with a manual SLR at its prime focus.

Digital imaging

With their highly sensitive CCD chips, digital cameras are capable of capturing celestial scenes in much shorter exposures than their conventional film counterparts.

Physically located where conventional film is placed in an ordinary camera, CCDs (charge coupled devices) are integral to an array of imaging equipment, including webcams, digital cameras, video cameras, and specialist astronomical cameras. CCDs are made up of a grid-like array of tiny electrical sensors called pixels which are activated by light falling upon them. The greater the number of pixels in a CCD chip, the better its ability to record fine detail. An entry-level digital camera may have a rating above 1 megapixels, its CCD chip having more than one million pixels. Mid-level cameras are rated above 4 megapixels, and advanced digital cameras may have highly sensitive chips up to 10 megapixels or more.

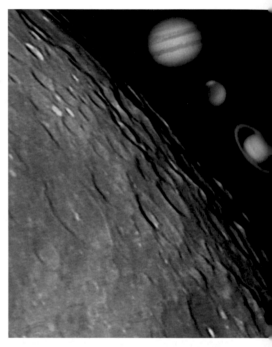

▲ Montage of lunar and planetary images, all to scale, taken with a webcam and 150mm refractor.

Still digital images

Digital imaging has a number of distinct advantages over conventional photography. A digital image can be viewed on the camera's electronic viewscreen immediately after it has been taken and, if unsatisfactory, it can be deleted from the camera's memory. In addition, digital cameras with a large capacity storage card can store many more images than conventional camera film. These images can be transferred to computer, enhanced and printed off at leisure—there is no need to take film down to the local processing store for results which may or may not be to your liking!

Digital cameras have their limitations when it comes to astrophotography. Unlike film cameras, they cannot be used to take exposures of low light level scenes for more than a few seconds without the resulting image being noticeably grainy, caused by electronic "noise" within the CCD.

▲ Philips ToUcam at prime focus of 127mm Maksutov, using x2 Barlow lens.

Pleasing wide-angle shots of bright lunar and planetary phenomena and bright aurorae can be obtained with a good digital camera. Digital cameras are most effective in capturing afocal images of the Moon and brighter planets through the telescope—these objects require shorter exposures; it is even possible to secure detailed images of the Moon with an undriven telescope. A telescopic lunar image taken with a superior digital camera will capture considerably more detail than an entry-level camera. Although both images may appear superficially identical, the one taken with the camera with the highest pixel rating can be zoomed-in on the computer screen to a much greater degree.

Webcams work wonders

Although the CCDs in webcams have a lower pixel rating than those in most basic digital cameras, webcams have become extremely popular with astroimagers seeking to produce high-resolution images of the Moon and brighter planets. By taking a short video of an astronomical object centered in the field of view, a series of dozens or even hundreds of individual frames are captured. When analyzed by a suitable computer program, each frame of the video can be graded according to its quality and the best ones processed to produce a single superior image.

◀ Philips ToUcam webcam with its lens removed, showing the tiny internal CCD chip.

want to know more?

Take it to the next level...

Go to...
- ▶ **Sketching the skies**—page 44
- ▶ **Conventional photography**—pages 46–7
- ▶ **Experiment**—page 47

Other sources
- ▶ **Magazines and books**
  essential reading to get you imaging
- ▶ **Manufacturer's guidelines**
  read the camera's instructions
- ▶ **Manufacturer's software**
  install all software with your camera
- ▶ **Imaging software**
  automatically process your images
- ▶ **Internet resources**
  lots of useful information on imaging

our cosmic

backyard

A tremendous variety of objects and events within the Solar System can be viewed by the vigilant stargazer, from amazing phenomena taking place high in the atmosphere to the Moon and planets beyond. Some of these are spectacular to the unaided eye, while others are visible when viewed through the telescope eyepiece.

Atmospheric effects

A number of easily visible astronomical phenomena take place within the atmosphere, a layer of gases just 372 miles (600km) deep that separates us from the harsh vacuum of outer space.

▲ An outstretched hand shields the eyes from direct sunlight to view the solar halo.

WATCH OUT!

Protection

Take care when viewing these phenomena by covering the Sun with your hand, or by obscuring it behind a nearby landmark to ensure safe, glare-free viewing.

Solar spectacles

Under a range of atmospheric conditions, the Sun produces a dazzling array of aerial effects. Everyone is familiar with the rainbow—a multicolored arc of light directly opposite the Sun in the sky. Rainbows are caused when sunlight is split (refracted) into the colors of the spectrum by water droplets in the air.

Sometimes the Sun is encircled by a halo—a luminous ring about 44° in diameter which often has a reddish inner border and a diffuse outer edge. Occasionally, prominent bright glowing patches can be seen about 22° on either side of the Sun. These are often referred to as sundogs, and they can appear brilliant and colorful. When seen together, a bright solar halo with sundogs on either side is a most impressive sight. Both haloes and sundogs are formed when sunlight is refracted by ice crystals high in the atmosphere (around the same height as commercial jet airliners), but each is caused by different shaped ice crystals.

Noctilucent cloud

Most weather takes place in the troposphere, the lower 9 miles (15km) of the atmosphere. But high above all the regular

▶ A vivid sundog.

▶ Noctilucent cloud.

clouds, at a height of around 53 miles (85km), floats a wispy layer of clouds composed of ice crystals. Being so high, these clouds can remain lit by sunlight long after the Earth directly beneath has been plunged into the shadow of night. This noctilucent cloud ("bright at night") can appear really dramatic when set against a dark evening sky.

The aurora

A constant stream of energetic particles (electrons and protons) is carried away from the Sun within the solar wind. These particles are channeled down into the atmosphere upon encountering the Earth's strong magnetic field, and when they hit gas atoms within the atmosphere, a luminous glow is produced. This fantastic light show is called the aurora, known as the *aurora borealis* in the northern hemisphere, and *aurora australis* in the southern hemisphere. Colors within aurorae are produced by different glowing gases (oxygen and nitrogen) between 40 and 120 miles (60–200km) high. Aurorae are produced in a region circling the magnetic poles, but on those occasions when there is a high bombardment of particles from the Sun, they can be seen from locations as far south as the

▼ Aurora Borealis.

▲ Aurora Borealis.

Mediterranean or as far north as Australia, and vivid ones can sometimes be bright enough to be seen from urban locations. Aurorae take on numerous forms, including broad and rayed arcs, flowing curtains and fabulous streaming coronae, all of which appear to change minute by minute as the activity progresses.

Meteors

A meteor is the glowing trail left when a small meteoroid—a bit of debris left in the wake of a comet—burns up in the Earth's atmosphere at a height between around 50 and 60 miles (80–100km). Though a fleeting phenomenon, the sudden flash and rapid flight of a meteor across the sky is exhilarating to view.

Annual showers

Every year the Earth passes through a number of meteoroid streams, producing annual meteor showers that occur at around the same date each year. Meteors produced by the annual showers appear to emanate from a well-defined radiant in the sky—an effect of perspective as the Earth passes through the stream—and the radiant is named after the constellation in which it lies. If you watch the area of sky near a radiant during the period of a major shower, you will be unlucky not to see at least one meteor within a quarter of an hour.

Despite the ominous appearance of the brightest meteors (including fireballs of exceptional brilliance), seen during some of the annual showers, they pose absolutely no threat to the stargazer. Meteoroids within the well known streams range in size from grapes to grains of sand, and they all burn up completely in the atmosphere. Those objects large and solid enough to survive a superheated descent through the atmosphere originated from

WATCH OUT!

Observing meteors

Observing is best done well away from overt light pollution and on a night when the Moon isn't high and near full phase. The nights can be chilly, so wrap up well.

OUR COSMIC BACKYARD

asteroids (and a rare few from the Moon and Mars), and they are not associated with the annual meteor showers.

The best meteor showers

Lyrids
Active from April 16 to 25. Maximum on April 21–22. Brilliant medium speed meteors.

Eta Aquarids
Active from April 21 to May 12. Maximum on May 5–6. Very fast meteors with persistent trains. The Eta Aquarids were produced by Halley's Comet.

Perseids
Active from July 23 to August 22. Maximum on August 12–13. One of the most popular of the annual meteor showers, the Perseids are very fast and often extremely bright.

Orionids
Active from October 15 to 29. Maximum on October 21. Bright, very fast meteor.

Leonids
Active from November 13 to 20 . Maximum on November 17–18. Fast meteors with persistent trains. Enhanced Leonid activity occurs from time to time.

Geminids
Active from December 6 to 19. Maximum on December 13–14. Very bright, intensely white meteors, slow moving.

Quadrantids
A shower that takes its name from a defunct constellation to the north of the tail stars of the Dipper are active between January 1 and 5; its maximum on Jan 4 can display rates of up to 120 meteors per hour.

▲ Shower meteors appear to emanate from a radiant. The Leonids' radiant lies within the prominent "sickle" of Leo.

▲ A very bright Leonid meteor. The constellation of Leo is on the left.

▶ Spectacular Sun

Although the Sun is an average-sized star—a "mere" 868,000 miles (1.4 million km) in diameter—it is far from mundane. The solar surface is a constant turmoil of activity, much of which is easily visible through a small telescope.

On the spot reactions

Inside the Sun, nuclear fusion converts hydrogen to helium. Each and every second, fusion converts a staggering four million tonnes of gas into energy. A powerful magnetic field is generated within the Sun, and disturbances within it cause sunspots to appear in the photosphere, the Sun's glowing surface. With a temperature of 5,500°C, the photosphere is up to 2,000°C hotter than a sunspot's interior; such a big contrast in temperature and brightness makes sunspots appear dark against the photosphere.

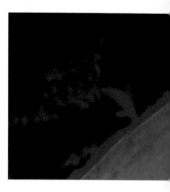

▲ Prominences can appear like large flames at the edges of the solar disc, but they can only be viewed through special H-alpha filters.

Solar cycle

Records of solar activity show that the Sun has an 11-year cycle, where the number of spots

WATCH OUT!

Observe the Sun in safety!
Never view the Sun through binoculars or a telescope—a fraction of a second of magnified sunlight can cause permanent blindness. The safest way to observe is to project the Sun's image onto a shielded smooth white card, but note that some telescopes and eyepieces have plastic parts that may melt if they are subjected to magnified sunlight. If your telescope has a finder, keep its lenses covered and don't attempt to locate the Sun with it. Don't leave your telescope unattended while observing the Sun, as less experienced people may be tempted to look at the Sun through it. More experienced observers use special solar filters that fit over the telescope aperture, and follow the instructions on their use to the letter. Never use small dark filters that fit on eyepieces, and never use any other household materials as a solar filter, as these are quite inadequate to protect your eyes from the potentially blinding solar radiation.

seen on the Sun's disc rises and falls. At solar minimum, when the Sun is at its least active, the disc can be spotless for several weeks. Sunspot maximum sees a daily proliferation of spots, with occasional giants that can be seen with the unaided eye through a proper solar filter. The next solar maximum takes place around 2011.

Observing the Sun

A sunspot at the western edge of the Sun will be carried across to the eastern edge in less than a fortnight. Most sunspots have a lifetime of less than one solar rotation (around 25 days at the Sun's equator), but big ones—those so huge that they could easily accommodate a dozen Earths—can survive several months. Sunspots have a fascinating structure which morphs from day to day. Their interior (called the umbra) is often dark and featureless, but the area surrounding it (the penumbra) is gray and often striated with a mass of radial lines. Large sunspot groups usually have a main header spot and a slightly smaller follower spot, among a number of others, all embedded within a larger penumbra.

Sunspots are the Sun's most conspicuous feature, but a large telescope will reveal a fine structure across the Sun's surface known as granulation, caused by a multitude of bubbling convection cells in the photosphere. Brighter areas called faculae are sometimes seen, often appearing most prominent toward the edge of the Sun, where the photosphere is less bright.

▼ Projecting the Sun's image onto a shielded smooth white card is by far the safest method of observing the Sun safely.

Sun shade (makes image of sun easier to see)

Protection screen (white card)

Full aperture Solar filter

Make sure your finder has its lens caps on!

◀ Direct view using full aperture solar filter.

▼ The Sun, showing a sizeable sunspot group.

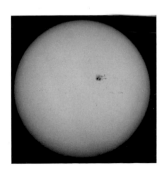

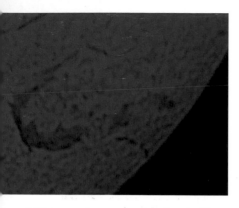

▲ When prominences are viewed against the Sun's disc in H-alpha light they appear as well-defined dusky swathes called filaments.

Seeing the Sun in a different light

A lot more activity on the Sun can be seen through special filters which block out all light except a band of red light called H-alpha. The chromosphere, a layer of the Sun's atmosphere just above the photosphere, emits H-alpha light, and within it large clouds of gas called prominences can be seen. Prominences can appear to change their appearance over just a few minutes, and they can develop into a variety of stunning forms. Like sunspots, prominences are associated with strong magnetic activity. H-alpha telescopes are expensive, and very few stargazers have ever had the pleasure of peering at the Sun through one.

Solar eclipses

It's an amazing coincidence that both the Sun and Moon are around a hundred times their own diameter away from the Earth, and they both appear to have around the same apparent size of half a degree as seen from the Earth. Occasionally the Moon passes in front of the Sun, blocking part or all of the Sun's light from reaching a portion of the Earth's surface. The darkest part of the Moon's shadow only just reaches the Earth under favorable circumstances, so total solar eclipses can only be seen along a small path as the Moon moves through space and the Earth revolves beneath the shadow. It is possible for some total eclipses to last for more than seven minutes, but most last for a much shorter time. A total eclipse is one of nature's most awesome spectacles. For a brief moment, the Sun is completely hidden as the viewer is plunged into darkness, it becomes chilly, and a spooky silence pervades the surroundings. Bright stars and planets become visible. The Moon's edge is punctuated by prominences, and the pearly

▼ The Sun, imaged in H-alpha light, showing granulation and prominences.

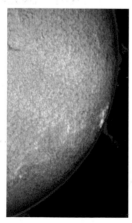

OUR COSMIC BACKYARD

◀ The totally eclipsed Sun imaged from Zimbabwe, June 21, 2001.

streamers of the corona, the Sun's outer atmosphere, become visible.

It is safe to view the Sun directly during totality, but never during an eclipse's partial phases. Either side of the line of totality is a broad zone where a partial eclipse is visible, and the further away from the central line, the smaller the partial eclipse will be. Sometimes the Moon's shadow doesn't quite reach the Earth, causing an annular eclipse, where the Moon briefly appears as a big black circle surrounded by a bright ring.

▼ The partially eclipsed Sun rises over Athens on May 31, 2003.

ECLIPSES

Future eclipses

Date	Type	Visible from
Oct 03, 2005	Annular	UK, Europe, Africa, S. Asia (Annular from Portugal, Spain, Libya, Sudan, Kenya)
Mar 29, 2006	Total	UK, Europe, Africa, W. Asia (Total from C. Africa, Turkey, Russia)
Sep 22, 2006	Annular	S. America, W. Africa, Antarctica (Annular from Guyana, Suriname, F. Guiana, S. Atlantic)
Mar 19, 2007	Partial	Asia, Alaska
Sep 11, 2007	Partial	S. America, Antarctica
Feb 07, 2008	Annular	Antarctica, E. Australia, N. Zealand (Annular from Antarctica)
Aug 01, 2008	Total	N. America, UK, Europe, Asia (Total from N. Canada, Greenland, Siberia, Mongolia, China)
Jan 26, 2009	Annular	S. Africa, Antarctica, SE Asia, Australia (Annular from S. India, Sumatra, Borneo)
Jul 22, 2009	Total	E. Asia, Pacific Ocean, Hawaii (Total from India, Nepal, China, C. Pacific)
Jan 15, 2010	Annular	Africa, Asia (Annular from C. Africa, India, China)
Jul 11, 2010	Total	S. America (Total from S. Pacific, Easter Island, Chile, Argentina)

► The Moon, our sister planet

For billions of years, the Moon has partnered the Earth in its orbit around the Sun. Its silent, doleful face has witnessed the rise and fall of the dinosaurs and the relatively recent arrival of humans on its big blue partner.

▲ This image of the Earth and Moon in space was taken by the NASA probe Galileo on its way to Jupiter in 1990.

No small satellite

Our lunar companion is instantly recognizable to anyone who has ever looked at the night sky. At a total of 2,155 miles (3,476 km) across, the Moon measures about the same breadth as the USA—around a quarter the diameter of the Earth. The Moon is our only known natural satellite, a sphere of rock whose size is only exceeded by four other satellites in the Solar System. Because of the Moon's relatively large size, the Earth and Moon have often been called a "double planet."

The "Big Whack"

It is incredible to imagine that the Moon might have been produced in a one-in-a-million chance collision between a small planet and the young Earth, but this is currently the most widely accepted theory to account for the Moon's origin. Other theories include the idea that the Moon was flung off the Earth by rapid rotation, that the Moon is a captured planet, or that the Moon formed out of a ring of debris left over from the Earth's formation, but none of these theories quite fits all the data we have about the Moon. The "Big Whack" theory envisions a Mars-sized planet colliding with the Earth. The impactor's heavy core joined with the material of the Earth, while the lighter molten mantle material of the impactor was

▲ Covered with vast lava flows and peppered with craters, the Moon's surface presents a permanent record of widespread volcanic activity and asteroid impacts

▼ The so-called "Big Whack" theory of the Moon's origin sees a violent birth of the Moon, when a Mars-sized planet struck the very young Earth and blasted off a huge molten mass into space, much of which coalesced to form the Moon.

mixed with that of the Earth and flung out into space, where much of it coalesced in orbit to form the Moon.

A turbulent history

Once the Moon had formed and its crust had solidified, the Moon experienced a phase of intense asteroid bombardment. Our own planet was similarly assaulted by the debris left over from the formation of the Solar System, but it has been obliterated by the incessant action of plate tectonics, deforming the crust, in combination with volcanic activity, erosion, and sedimentation, processes that do their best to hide the topography of past epochs. But the Moon's crust has long been solid and immovable, and etched deep into its face can be seen the signs of both volcanic activity and asteroid impacts. Most of this activity took place billions of years ago, long before dinosaurs had eked out a clawhold on the Earth. Yet we can clearly see many of these ancient features through binoculars and telescopes. Some really ancient craters, formed billions of years ago, are so well-preserved that the uninformed observer might imagine that they had been formed in recent times!

With its relatively small mass and low gravity, unlike the Earth, the Moon never managed to hold onto a substantial atmosphere, nor did water ever gush across its surface. No lunar life of any kind has ever evolved on its surface, and the Moon has remained utterly sterile since its formation, 4.6 billion years ago.

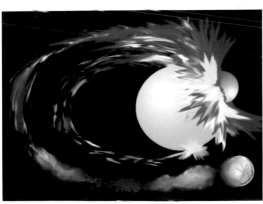

▲ The lunar occultation of Saturn, April 2002.

▲ Since the bright star Regulus in Leo lies near the ecliptic, it is prone to occasional occultations by the Moon.

▲ The total lunar eclipse of October 2004.

The Moon in space

At an average distance of 238,328 miles, the Moon revolves around the Earth in a near-circular orbit once every 29.5 days. As it orbits, the Moon keeps one face turned toward the Earth, keeping most of its far-side perpetually hidden from our view. The Moon's apparent size is about half a degree across—so small that it can be covered by the tip of your little finger. When the full Moon appears near the horizon, it can seem to be much larger than usual, but this is just an illusion caused by the way we perceive the shape of the sky and the objects in it.

The Moon's phases

As the Moon orbits the Earth, its angle of illumination by the Sun slowly changes, and the shape of the illuminated portion of the Moon (its phase) changes from a thin crescent, through "first quarter" (half Moon), to full Moon, "last quarter" (half Moon), thin crescent, and back to new Moon. At its full phase, the Moon is opposite the Sun in the sky and appears completely lit. At new Moon, the Moon lies between the Sun and the Earth and is not illuminated at all from our perspective. This cycle of lunar phases repeats itself every 29.5 days.

Lunar eclipses

Sometimes the full Moon moves through the shadow cast by the Earth into space, and experiences an eclipse. On such occasions the Moon doesn't completely vanish from sight, even if it happens to plunge deep into the Earth's shadow, because a certain amount of sunlight is bent around the edge of the Earth by the atmosphere and reaches the Moon's surface. This bent light is, however, strongly red in color, and as a result the totally eclipsed Moon appears reddish in color. Its hue varies from eclipse to

Earthshine

A phenomenon called "Earthshine"—the faint illumination of the Moon's dark side—is most obvious when the Moon is a slender crescent set in a dark sky. This delightful spectacle is caused by sunlight being reflected from the Earth onto the Moon's dark side. Through binoculars it is even possible to identify some of the larger surface features that are illuminated by the Earthshine alone.

eclipse, ranging from bright orange to deep ruddy brown. A lunar eclipse can last for several hours from the point that it first enters the Earth's shadow to the moment it leaves the shadow. Lunar eclipses are best viewed through a pair of steadily held binoculars—the viewer gets the wonderful impression of the Moon floating in space at totality, with the darkened Moon set among a star field that might normally be invisible at full Moon.

Hide and seek

During its course through the skies, the Moon frequently passes in front of stars and hides them for a while. These events are known as "occultations," and they can be fascinating to view through the telescope. Often, a star will appear to switch off as the Moon's edge covers it—a dramatic event that never fails to astonish the observer by its sheer suddenness. Occasionally a planet is occulted by the Moon, and such events are eagerly anticipated, as the occultation of a large object such as Venus, Jupiter, or Saturn can take many tens of seconds to complete; to have the Moon and a planet in the

same high magnification field of view provides a wonderful visual delight. Dates and times of bright lunar occultations are printed in several publications.

◀ The phenomenon of Earthshine on a waxing crescent Moon.

▲ A view of the sunrise terminator of the first quarter Moon. Craters near the terminator appear full of shadow and incredibly deep—an illusion caused by the low angle of illumination. The large flat-floored crater at upper center is Ptolemaeus, 95 miles across.

▲ Mare Imbrium is the Moon's largest circular sea, and is magnificent to view at first quarter phase.

Observing the Moon

Being so close to the Earth, the Moon appears large and bright, and the biggest of its surface features can be seen with the unaided eye alone. It's thrilling to view the Moon through a pair of steadily-held binoculars. A number of large dark flattish areas surrounded by mountain ranges are the most obvious lunar features. These are the maria (Latin: seas)—vast expanses of lava that flowed and solidified several billion years ago. Maria are generally circular because they occupy the floors of huge asteroid impact basins. Binoculars will also show hundreds of smaller impact craters, and their forms vary depending on their age, size, and how much other things (such as volcanic activity and overlying impacts) have altered their appearance since they were formed. The youngest craters are sharp and bright, and many are surrounded by prominent rays of ejected material which can stretch across the lunar surface for hundreds of miles.

Craters abound on the Moon, especially in the southern uplands. Some ancient craters are highly eroded and are only visible at low angles of illumination. Other craters show magnificent structure in their walls and impressive mountains rising at their centers, while a number of large craters have been largely flooded with lava.

Telescopic lunar delights

A telescope will show much more on the lunar surface. Inside the maria are bays and ghost craters—old impact craters whose walls have been breached by lava flows and partly filled by them. Isolated mountains can be found in many of the maria; these are remnants of the basin's inner walls which poke out above the lava flows. Close observation reveals low rounded hills called domes, which are long-extinct volcanoes, complete with summit vents. Low ridges run across the maria, wrinkles formed after the lava flows had ceased. These features require a low illumination to be seen.

Large and impressive mountain ranges border many of the maria, most notably around the Moon's largest, Mare Imbrium, whose edge is marked by the Jura mountains, the Alps, Apennines, and Carpathians (each named after their terrestrial counterparts).

Long, curving fault valleys called arcuate rilles cut through the Moon's crust around the edges of some of the maria, and the floors of some individual craters are cut through by straight clefts called linear rilles.

Activity on the Moon?

Transient lunar phenomena (TLP) are rarely observed localized colored glows, obscurations, and brief flashes. There is little hard scientific evidence to support the existence of TLP. It is possible that glows and obscurations are caused by lunar degassing and electrical activity, and flashes might be caused by the occasional impact of large meteorites.

▼ A waxing gibbous Moon, nine days after new Moon. Near the northern terminator is the large Bay of Rainbows, and the Sea of Clouds can be seen towards the south, with the large crater Gassendi perched on its northern shore.

◀ Full Moon is the best time to view the Moon's ray craters. Half a dozen prominent ray craters can be viewed through binoculars, notably those surrounding Tycho in the south and Copernicus towards the west.

► The inner planets

**Both Mercury and Venus orbit the Sun closer than the Earth.
They never appear to stray very far from the Sun, and the
two planets can often be seen with the unaided eye shining
in the dawn or dusk skies.**

Mercury, closest rock to the Sun

Mercury is the most elusive of the naked eye planets. It whizzes around
the Sun once every 88 days, and is only favorably placed for viewing for
around a week or two when at its greatest angle from the Sun (called
"maximum elongation"). City dwellers may find Mercury a difficult planet to
spot because it always appears rather low down near the sunrise or
sunset horizon. To be able to locate Mercury, you need a reasonably clear
horizon, and the planet must be high enough for its light to shine through
the near-horizon atmospheric murk. The best chances to spot Mercury
occur when its maximum elongation takes place in the evening skies of
spring, when it appears above the western horizon after sunset, or the
morning skies of autumn, when it can be seen above the eastern horizon
before sunrise.

Mercury can appear almost as bright as
Sirius, the sky's brightest star, and it shines with
a rosy hue. Through a telescope, this little world
—a bare rock around twice the diameter of the
Moon—doesn't show a great deal on its small
disk, though its phase can be discerned at a
high magnification. It was only when the Mariner
10 spaceprobe imaged the planet more than
30 years ago that we first got to know that its
surface is very heavily cratered, like the Moon's
highlands.

▼ Mercury presents a tiny
disk, upon which features
may occasionally be
glimpsed.

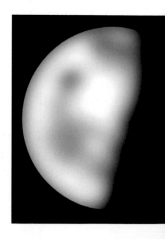

Venus, the hell planet

Venus can be a conspicuous object in the
evening or morning skies, dazzling viewers with
its brilliance. Venus' orbit around the Sun takes
a leisurely 225 days, which means that there
can be a maximum of two elongations from

the Sun each year. At maximum elongations, Venus reaches a respectable angle of 45° from the Sun, enabling it to appear high in the sky for several hours after sunset or before sunrise. Like Mercury, Venus is at its highest above the sunset horizon when its maximum eastern elongations take place during the spring; it is highest above the sunrise horizon when its maximum western elongations occur during the fall.

Seen in a dark sky, Venus is a bright object that shines with an intense white color. Around the same size as the Earth, Venus is swathed in clouds so thick that its surface can never be seen through the telescope. It was once thought that conditions on Venus might be suitable for life to flourish, but that notion changed utterly when spaceprobes went there and landed on its surface. Venus, named after the beautiful goddess of love, is the nearest planetary environment in the Solar System to old fashioned ideas about hell! Clouds of sulphuric acid drift through an atmosphere of carbon dioxide. On Venus' surface the atmospheric pressure is higher than in a pressure cooker, and temperatures are far hotter than a kitchen oven.

Seen through a telescope, some Venusian cloud features are occasionally visible, but its phases are obvious, varying from a small gibbous disk to a sizeable crescent during each apparition.

On rare occasions Venus passes directly between the Sun and the Earth, and it appears as a black spot that travels across the Sun in a few hours. The last transit of Venus took place in June 2004, providing a spectacle that was widely observed.

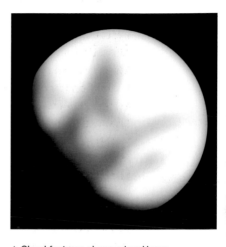

▲ Cloud features observed on Venus.

WATCH OUT!

Easy to mistake
Venus is sometimes mistaken for a UFO by inexperienced and experienced observers such as pilots; it can be hard to identify!

► Mars attracts

Of all the planets in the Solar System, the red planet Mars has attracted the most attention from both science and popular culture, because it has long been known to be similar to the Earth in many ways.

MUST KNOW

Life on Mars?
Spaceprobes have detected minerals that could only have formed in water, which flowed across the surface in the remote past when Mars was warmer.

▼ Mars, observed on July 10, July 20, and August 30, 2003, the latter observation being made just two days after the planet's historic closest approach to the Earth. In the observation at left is Solis Lacus, the "Eye of Mars." In the observation at center, the polar cap has a large split in it, and at right is the planet's most prominent feature, Syrtis Major.

Orbiting further from the Sun than the Earth, Mars is just over half the Earth's size. Its day is only 37 minutes longer than our own, and its axial tilt is similar to the Earth's, meaning that the planet experiences seasons. For centuries, astronomers have watched in fascination as Mars' polar ice caps vary with the seasons, suggesting that the ice melts when the planet warms up in the Martian summertime and refreezes during the winter. Seasonal changes affect the dusky markings on Mars, too, as some areas appear to broaden and darken, while others fade. Yet the markings always appear to return to their original configuration, regardless of any temporary changes they might experience. Mars has long been known to have an atmosphere, too, as occasional bright clouds and large dust storms have been observed through the telescope. Some dust storms have been so severe as to completely obscure all the markings on the planet.

A closer look

Mars is colder than the Earth, and it has a thin atmosphere of carbon dioxide. Bright clouds of

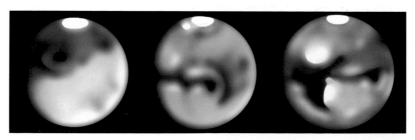

ice crystals occasionally form, and the wind whips up anything from small dust devils to occasional seasonal dust storms which obscure large swathes of the planet. The wind is responsible for the changing appearance of the dusky features on Mars, as the planet's dark surface material is temporarily hidden by dust or exposed as the dust is blown away.

Rocky deserts cover much of Mars, colored red because they are literally "rusty" with the mineral iron oxide. Craters predominate in the planet's southern hemisphere, while large portions of the northern hemisphere consist of smooth rolling plains.

An impressive collection of extinct volcanoes dominates the Tharsis region. Of these, Mount Olympus, the Solar System's biggest volcano, is 310 miles wide and three times higher than Mount Everest. Clouds over Mount Olympus can sometimes be seen through backyard telescopes. East of Tharsis lies the incredible Mariner Valley, a giant rift more than 1,860 miles long, 372 miles across and 5 miles deep in places. Some dark parts along the valley floor can be seen through a telescope.

Observing Mars

Most of the time Mars is too small for much detail to be seen, but for a few months every couple of years the planet is close enough for its ice caps and its dusky deserts to be easily made out. Most of Mars' dark markings are located in the southern hemisphere. The wedge-shaped plain of Syrtis Major, along with Hellas, a bright area to its south, are familiar to all Mars observers. On the other side of the planet, the dark eye-shaped Solis Lacus can appear striking. Mars' two satellites, Phobos and Deimos, are far too small and dim to be seen by casual backyard observers.

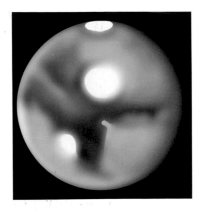

▲ Telescopic observation of Mars. The V-shaped Syrtis Major is prominent, so too is the bright Hellas basin and the south polar icecap.

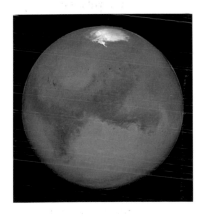

▲ Mars, imaged in July 2003 a few weeks before its closest ever approach to Earth.

▶ Jupiter, by Jove!

A dozen times broader than the Earth, Jupiter is the Solar System's biggest planet—a rapidly spinning ball of gas whose cloud features are fascinating to view through backyard telescopes. Its four largest moons shine brightly and are easy to spot.

▲ Jupiter and its Great Red Spot (near left edge of Jupiter).

The undisputed king of planets, Jupiter is more voluminous than all the Solar System's other planets, satellites, and asteroids put together. Jupiter has a fantastically fast rate of spin, revolving once on its axis in less than ten hours. Since the planet is made up of gas (mainly hydrogen and helium), its rapid spin produces marked centrifugal bulging at its equator.

Turbulent clouds

Jupiter has no solid surface, and its upper atmosphere is in constant turmoil, so no permanent features exist on the planet. Banding of the planet's dark belts and light zones, produced by the planet's rapid spin, can easily be seen through a small telescope. The belts and zones vary in intensity from year to year, but the most prominent are usually the North and South Equatorial Belts. Features within the cloud belts and zones change from week to week, as spots, ovals, and festoons develop, drift in longitude, interact with one another, and fade away. Features remain within their own belt or zone, and they never drift in latitude.

Astronomers have kept a more or less constant telescopic watch on developments on Jupiter for more than a century. Perhaps the nearest thing that Jupiter has to a "permanent" feature is the famous Great Red Spot, a giant anticyclone that could easily swallow the Earth. Having wandered around the planet's South Tropical Zone since at least the mid-19th century, the Great Red Spot varies in intensity from year to year, ranging from a barely discernible gray smudge to a sharply defined brick red oval which is easily visible through small telescopes. Other bright Jovian atmospheric spots have lasted for many years.

Satellite tracking

Sometimes the satellites drift directly across Jupiter and you can follow the black shadows they cast on the planet.

▼ The Great Red Spot is a giant long-lived anticyclone in Jupiter's South Tropical Zone.

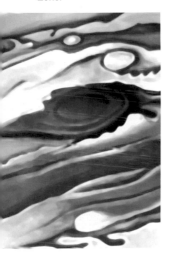

▼ Jupiter and two of its Galilean moons.

Observing Jupiter

Jupiter orbits the Sun every 12 years and moves slowly along the zodiac. It appears as a brilliant white star, second only to Venus (and occasionally Mars at its brightest), so it isn't difficult to locate with the unaided eye.

Binoculars will reveal Jupiter's four largest satellites—Io, Europa, Ganymede, and Callisto— as bright points of light.

Jupiter appears as a flattened disk through a small telescope at low magnification. A high magnification using at least a 100mm telescope is required to see much detail within the planet's dark belts and bright zones. Features appear near Jupiter's western edge, and rotation carries them to the central meridian—the imaginary line connecting a planet's poles—in just a couple of hours. Within a few more hours they are out of sight beyond the other edge of the disk.

Usually there is a good deal of activity in the equatorial regions, with festoons whipping off the south edge of the North Equatorial Belt, skirting around bright areas within the Equatorial Zone. Nestled in the South Tropical Zone, producing a marked indentation in the south edge of the South Equatorial Belt, the Great Red Spot is usually visible when facing the Earth, though its color may not be obvious. Small spots and ovals are often visible within Jupiter's atmosphere. You can sketch the planet and trace the movement of these features over a period of weeks.

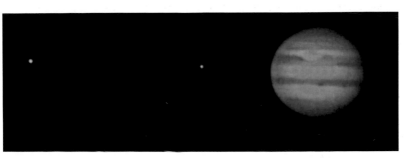

► Ringworld Saturn

On first observing the ringed planet Saturn, its sheer beauty both surprises and awes the viewer. With its wonderful ring system and array of moons, Saturn never fails to provide a great visual delight to the telescopic viewer.

Saturn orbits the Sun way beyond Jupiter, completing a circuit in just over 29 years. Like Jupiter, Saturn is a giant, rapidly spinning ball of gas which is noticeably flattened at the poles. It is far less dense than Jupiter—less dense than water, in fact—so if you could find a big enough pool of water, Saturn would float in it!

Saturn's atmosphere displays dusky bands and brighter zones parallel to the equator, but they are less pronounced than those of Jupiter. There is much less obvious activity within Saturn's atmosphere too, and localized features that appear distinct enough to be seen through the telescope, such as dark spots or bright ovals, are rare. Large bright spots billowed up in Saturn's atmosphere in 1933, 1960, and 1990, but these were short-lived and faded after a few weeks.

▲ These images, taken by the Hubble Space Telescope between 1996 and 2000, shows how the tilt of the planet changes over time, in this case the rings broadening as the planet's north polar region comes into view.

Remarkable rings

Saturn's ring system is exactly in line with the planet's equator. It is tilted to the plane of its orbit around the Sun, and our view of the rings changes year after year. From being at their widest, the rings appear close to a thin line in around seven years; after another seven years, the rings once more appear wide open, but with Saturn's other pole presented to us.

Saturn's ring system has three main components. Ring A, the outermost ring, is slightly darker than Ring B, the middle main ring, and they are separated by a narrow gap called the Cassini Division. Both the A and B rings appear uniform and opaque, and they cast a dark shadow onto Saturn. The innermost ring, Ring C, is difficult to observe because it is faint and somewhat transparent, qualities that inspired its unofficial name of the "Crepe Ring."

Saturn under scrutiny

Saturn shines as a bright yellowish star which is easy to identify in the night sky. Binoculars won't usually show the rings clearly, but Titan can be discerned. Saturn is beautiful through any telescope. At a low magnification, the planet and its satellites can be seen in the same field of view. Steady seeing is required for a really good high magnification view of Saturn. Depending on the angle at which we view the planet, the shadow of the rings on the globe and the globe's shadow on the rings can clearly be seen.

▲ Imaged from an angle never attained from the Earth, this amazing view of Saturn and its rings was taken by the Cassini spaceprobe in December 2004.

▶ Observation of Saturn and its satellites in 1986.

▶ The outer limits

**Saturn was long considered to be the last planetary outpost
of the Solar System. That changed in 1781 when Uranus was
discovered, doubling the size of the Solar System overnight.**

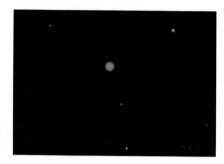

▲ Telescopic observation of Uranus.

Uranus circles the Sun at a vast distance
beyond Saturn—around 20 times the distance of
the Earth from the Sun, and so far away that it
takes more than 84 years to complete one orbit.
Uranus is a giant ball of gas, with a diameter
more than four times that of the Earth.

If you know exactly where to look, Uranus can
be seen with a keen naked eye, if the sky is clear
and dark. It is easy to spot with binoculars, but to
be able to discern its disk a high magnification is
required. Through a high-power eyepiece,
Uranus appears as a tiny featureless disk with a
pale greenish hue. Little
of interest was
revealed by the
Voyager 2 spaceprobe
as it flew by the planet
in February 1986,
except for a dusky
polar region and a few
vague bands and
spots. It wasn't the
planet itself that proved
to be the highlight of
Voyager 2's flyby
through the Uranian
system, but its retinue

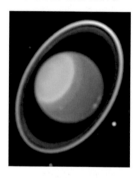

▲ Uranus, encircled by its
narrow ring system and
eight of its small satellites,
imaged by the Hubble
Space Telescope.

of small satellites, all of which are too faint to be
seen through backyard telescopes. Uranus also
has a ring system, but this is far less spectacular
than that of Saturn, and so faint that it can only be
seen on images taken through big telescopes.

Neptune is the smallest of the Solar
System's gas giants. With an orbital period of

74

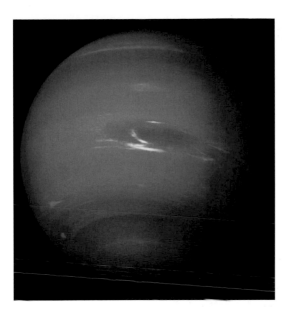

▲ Neptune, showing cloud features including the "Great Dark Spot" (imaged by Voyager 2).

▼ During eclipses of Pluto by its satellite Charon from 1985–1990, the planet's brightness varied, allowing this chart to be made of its surface. Pluto appears a brownish color.

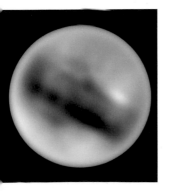

164 years, Neptune lies at the staggering distance of 2.8 billion miles from the Sun—so distant that sunlight takes about four hours to reach it. It takes 165 years to orbit the Sun. Neptune currently resides deep in Capricornus, moving into neighboring Aquarius in 2010. Observers find Neptune more challenging to locate than Uranus. Too faint to be seen with the unaided eye, it can be spotted as a faint star-like point through binoculars. Neptune's disk is tiny, and to resolve it requires at least a 100mm telescope with a minimum magnification of x200. Observers might just be able to discern a slight bluish tinge, but obvious features are not visible through small telescopes.

Pluto is the smallest of the Solar System's nine "official" planets. Measuring just two-thirds the size of the Moon, this tiny icy world lies way beyond Neptune. Images taken by the Hubble Space Telescope have revealed a vague patchwork of light and dark areas on Pluto, but what these might be is anyone's guess. Most amateur astronomers have never observed Pluto—it is far too faint to be seen in anything smaller than a 250mm telescope, and even when seen through a really large telescope at maximum magnification, it just looks like a faint star. Still, the thrill of actually seeing Pluto captivates many stargazers, worth every effort made in the hunt.

OUR COSMIC BACKYARD

Interplanetary debris

Comets and asteroids have had a bad press over the ages. In ancient times, people imagined that comets were harbingers of doom. More recently, movies have portrayed these cosmic itinerants as potential threats to human civilization.

MUST KNOW

Comet tail

Gases are blown by the solar wind (energetic particles streaming out from the Sun), forming a straight tail that points directly away from the Sun.

▼ The cratered nucleus of Comet Wild 2, imaged by NASA's Stardust probe as it flew by in January 2004. This composite image also shows the jets, gas, and dust surrounding the nucleus.

Comets, celestial ghosts

When far away from the Sun, chilling out in the interplanetary deep freeze, a comet is a pretty unimpressive sight—a solid ball of ices mixed with dirt and rock just a few miles across. Known as the nucleus, this "dirty snowball" is made up of material left over from the formation of the Solar System.

As it approaches the Sun, the outside of the nucleus warms up and the ices sublimate, turning into gas. Streaming off the nucleus, the gas carries with it grains of dust, forming a *coma*, which envelops the nucleus in a haze of tens of thousands of miles across. Pulled by the Sun's gravity, the comet speeds up as it heads through the inner Solar System. Often a comet will develop a prominent tail. A comet's tail comprises two distinct components, one made up of gases and the other made up of dust. Reflecting sunlight, the curving dust tail is usually much brighter than the gas tail, and often shows considerable structure through binoculars. Although a bright comet's tail looks substantial, and can stretch for many tens of millions of miles, it is all cosmic showmanship. It has been said that a

▲ Binocular observation of comet C/2001 Q4 NEAT near the star cluster Praesepe on May 15, 2004.

comet is the nearest thing to nothing that can be something.

Snowball circuits

A number of comets follow well-known orbits within the Solar System, the most famous being Halley's Comet, whose orbit takes it from the realms beyond Neptune to the inner Solar System every 76 years. Some, such as 1997's spectacular Hale-Bopp, have orbital periods of thousands of years. Another type of comet has never been warmed by the Sun before, having spent all its life at the very limits of the Sun's influence in a region known as the Oort Cloud. Thought to be a vast shell made up of billions of cometary nuclei, the Oort Cloud extends halfway to the nearest stars.

Observing comets

Most newly discovered comets never develop into anything more spectacular than dim fuzzballs that can only be glimpsed through binoculars or telescopes. Now and again—say, once or twice a decade—a comet bright enough to be viewed with the unaided eye appears.

Binoculars are ideal for observations. At high magnification through a telescope the comet's tiny nucleus appears as a bright dot, sometimes surrounded by intricate structure such as jets, arcs, and concentric shells of reflective dust.

▶The magnificent Comet Hale-Bopp, imaged in April 1997.

Asteroids, vermin of the skies

▲ 285 miles in diameter, Vesta is the third largest asteroid in the Solar System. This image was taken by the Hubble Space Telescope in 1997.

When spacecraft such as Voyager 2 sped through the Sun's main asteroid belt between Mars and Jupiter, mission controllers had no concerns that their billion dollar probes would accidentally bump into anything. While there are hundreds of thousands of asteroids within the main asteroid belt, they are spread throughout an enormous volume of space. If you were standing on an asteroid, the nearest other asteroid would likely be a distant, dim, starlike point of light.

From the Earth, none of the asteroids are bright enough to be seen with the unaided eye. So many asteroids appear on long exposure professional photographs that they have been referred to as "the vermin of the skies." At the beginning of the 21st century, hundreds of thousands of minor planets have been identified. Of those orbiting in the main asteroid belt, Ceres is the largest, with a diameter of nearly 620 miles, and there are 26 asteroids larger than 124 miles. Surprisingly, if all the known asteroids were gathered together they would only form an object around half the size of our own Moon.

In addition to the minor planets within the main asteroid belt, there are other groups of asteroids elsewhere in the Solar System. Near-Earth asteroids are those whose orbit takes them close to that of the Earth, including Potentially Hazardous Asteroids—objects that could possibly collide with us at some point in the distant future. An interesting group called the Trojan asteroids are clustered at points far preceding and following Jupiter in

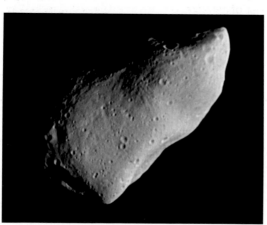

◄ Gaspra is a 12 mile-wide rock, one of thousands of similar objects in the asteroid belt. This image was taken in 1991 during the Galileo spaceprobe flyby.

its orbit around the Sun, locked there in a gravitational resonance.

It is thought that many of the asteroids in the main belt were once part of larger bodies that broke up by mutual collisions. These parent bodies, formed early in the history of the Solar System, grew hot inside and developed a core, mantle, and crust. Most of the meteorites found on the Earth are fragments of asteroids,

▲ This 4-inch diameter meteorite is thought to be a fragment of the crust of asteroid Vesta. Most meteorites are fragments of asteroids.

◀ Asteroid Ida, 10 miles wide, imaged in 1993 by the Galileo spaceprobe. At right is Ida's tiny 770-yard diameter satellite, named Dactyl.

and by studying their composition we can tell what materials the original parent bodies were made of.

Observing asteroids

Although the minor planet Vesta can sometimes reach the border of naked eye visibility (given excellent eyesight and a crystal clear sky), asteroids are far too faint to be seen without optical aid. Binoculars will reveal around a dozen of the larger main belt asteroids at their brightest, when they are located opposite to the Sun. To identify an asteroid, a good chart showing the asteroid's position among the stars is required. Regular observation over a period of days and weeks will disclose the asteroid's slow path, and it can be a rewarding exercise to produce a plot based on your own observations.

want to know more?

Take it to the next level...

Go to...
▶ Sketching the skies—page 44
▶ Conventional photography—pages 46–7
▶ Experiment—page 47

Other sources
▶ Magazines
  keep updated on planetary phenomena
▶ Join a society
  share your observations and improve
▶ Planetarium software
  plan your Solar System observations
▶ Print your own charts
  to help locate faint asteroids and comets
▶ Internet resources
  information about current events

deep

space

Our Solar System is
embedded within a galaxy
full of stars, star clusters, and
nebulae. Scattered beyond
the Milky Way, at unimaginable
distances in space and time,
lie countless more galaxies,
some like our own, others
vastly different in appearance.
Hundreds of deep sky objects
can be viewed through the
telescope eyepiece.

▶ A galaxy of stars

Born within vast clouds of dust and gas inside galaxies, stars shine by nuclear fusion, as atoms are squeezed together to form new elements and energy is given off in the process.

Although the stars may look superficially similar with the unaided eye, colors in the brighter stars can be discerned. Binoculars will reveal great variation in the color of the stars, from reds through whites, to steely blues. A star's color tells us how hot its surface is; redder stars are cooler, while bluer stars are hotter.

Nebulous clusters

A number of star-forming nebulae within our own Milky Way are visible through binoculars. About 1,500 light years away—virtually on our galactic doorstep—lies the Orion Nebula, the biggest and brightest nebula in the sky, a stellar maternity ward; it is visible as a misty patch below Orion's Belt. Just a short hop to its northwest is the young open star cluster of the Pleiades, a group of dozens of stellar pre-schoolers just starting their journey around the galaxy.

As a nebula contracts under its own gravity, its material heats up. Parts may grow so hot that nuclear reactions are ignited and a star is born. It takes a startup mass of about eight per cent that of the Sun to create a true star. Lightweight stars like these are called red dwarfs and have a "cool" surface temperature of around 3,000° K. Proxima Centauri, the nearest star (just over four light years away), is a red dwarf, though it's too faint to be seen with the unaided eye. Red dwarfs are the longest-lived of all the stars.

Our own Sun, a yellow G-type star, is 1,000 times brighter than Proxima Centauri and almost twice as hot. The Sun is similar to our near neighbor Alpha Centauri. Sirius, the brightest star in the skies, is around 80 times brighter than the Sun, more than three times as massive, and shines with a blue-white light. Sirius appears so bright because it is just 8.6 light years distant.

Twilight of the stars

There comes a time in the life of every star when its primary fuel, the hydrogen concentrated in its core, begins to deplete. When the star begins to burn hydrogen in its outer layers, it expands and turns into a red giant.

Near the end of its spell as a red giant, a star's outer layers expand out, forming a planetary nebula. The star at the center has become a white dwarf.

Stars that are far bigger than the Sun become red supergiants. Supergiants are so massive that their nuclear reactions go into overdrive. They become unstable, collapsing under their own gravity, its core crushed into a super-dense object known as a neutron star. Infalling material rebounds from the neutron star's surface, and the star's outer layers are blasted into space in a cata-strophic explosion called a supernova. Supernovas can briefly outshine the combined light of all the stars in its home galaxy.

▶ The Hertzsprung-Russell diagram compares the color and luminosity of stars with their temperature and color (spectral type). When plotted on the graph, most stars lie along a band called the "main sequence"—from dim red dwarfs at lower right, to massive, ultrabright, hot, blue stars at upper left. In addition, there are brilliant red giants and supergiant stars above the main sequence, while below the main sequence are low-luminosity white dwarfs.

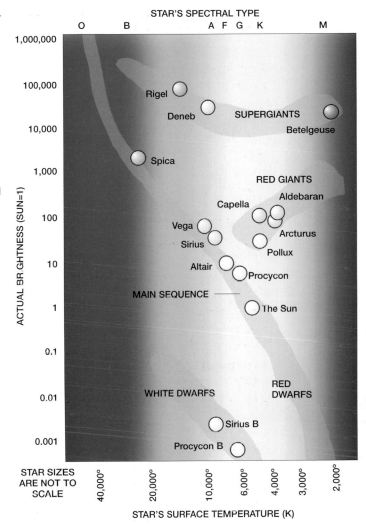

STARS OF THE MAIN SEQUENCE

Star type	Color	Surface temperature	Ave. mass (Sun = 1)	Ave. diameter (Sun = 1)	Ave. brightness (Sun = 1)	Main sequence example
O	Blue	25,000° K+	60	15	1,400,000	Zeta Puppis
B	Blue	11,000° - 25,000° K	18	7	20,000	Spica
A	Blue	7,500° - 11,000° K	3.2	2.5	80	Sirius
F	Blue to white	6,000° - 7,500° K	1.7	1.7	1.3	Procycon
G	White to yellow	5,000° - 6,000° K	1.1	1.1	1.1	Sun
K	Orange to red	3,500° - 5,000° K	0.8	0.8	0.9	Pollux
M	Red	below 3,500° K	0.4	0.4	0.04	Proxoma Centauri

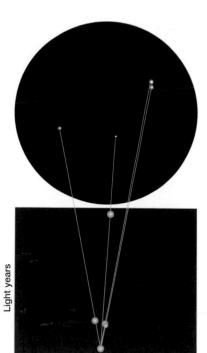

Light years

▲ The Mizar system, observed through a small telescope (top). Beneath, the relative distances of each component from the Sun are shown to linear scale, but angles are exaggerated.

Star partners

Many stars which appear to be close together are simply optical doubles, their apparent proximity to one another being caused by the angle at which we view them. Optical doubles consist of two stars with an immense physical separation, so distant that they are not gravitationally influenced by one another. True double and multiple stars—those relatively close to each other and orbiting their common center of gravity—abound through the galaxy. The cosmos evidently enjoys making them, since around half of all known stars are double or multiple systems.

Victorian astronomers took particular pleasure in identifying double stars, measuring their separations and noting their colors. Modern stargazers can take great pleasure from observing them—especially doubles with prominent colors or color contrasts—in addition to testing the resolving power of their optics and, in some cases, pushing their observing skills to the limits. These days, double star observation is a less popular area of professional astronomy.

Mizar, the second star from the end of Ursa Major's tail, is perhaps the most famous naked eye double star in existence. Keen-sighted folk will discern its fainter partner, Alcor, just a fifth of a degree to its northeast without optical aid. Known to the English as "Jack and his wagon" in years gone by, the pair are relatively close together in space, separated by just three light years. A small telescope will reveal that Mizar has a much closer partner than Alcor, lying just 14 arcseconds away. Interestingly, this pair was the first double star ever discovered through the telescope, way back in 1650. To add to the complexity of the system, both Mizar and its close partner are themselves binary stars (these cannot be

resolved through the eyepiece), making Mizar a quadruple system. The close binary pair of Mizar orbits its common center of gravity once every 5,000 years, while far out Alcor may take up to a million years to make one circuit of the Mizar system.

Epsilon Lyrae, another popular bright double star, can be separated without optical aid by those with really good eyesight. Each of the stars in this binary has its own partner, making it a "double-double" star, each component of which can be resolved through a 100mm telescope at a high magnification. It is also worth mentioning Theta Orionis (the Trapezium). Located at the heart of the Orion Nebula, this spectacular group of young, hot stars illuminates their gaseous neighborhood. Small telescopes show its four brighter components.

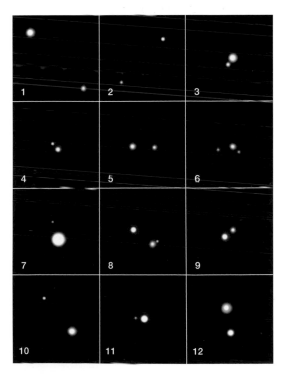

◀ Dozens of lovely colored double stars can be viewed through a small telescope. The stars featured here are some of the most beautiful examples. Colors have been exaggerated for clarity.

1. Beta Cygni (Albireo). Charts 6 & 7.
2. Eta Persei. Charts 2 &3.
3. Epsilon Bootis. Charts 5 & 11.
4. Xi Bootis. Charts 5 & 11.
5. Gamma Delphini. Chart 13.
6. Iota Cassiopeiae. Charts 1 &2.
7. Beta Orionis (Rigol). Chart 9.
8. Upsilon Andromedae. Chart 2.
9. Alpha Herculis. Chart 12.
10. Alpha Canum Venaticorum (Cor Caroli). Charts 4 & 5.
11. Alpha Scorpii (Antares). Charts 17 & 18.
12. Beta Scorpii. Charts 17 & 18.

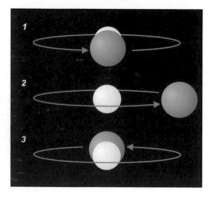

▲ Light curve of Algol, a bright eclipsing binary star. A smaller dip in brightness around maximum takes place when the secondary star is itself eclipsed by Algol.

Variable stars

Stargazers in ancient times knew that some stars varied in brightness over a period of time. The brightest and best-known variable star is Beta Persei (Algol, the "Demon"). Algol is a type of variable known as an eclipsing binary, its change in brightness caused when its light is temporarily obscured by a larger, dimmer star in orbit around it which produces regular eclipses. Unfailingly, every 2.87 days, Algol's brightness falls from magnitude 2.1 to 3.4; the eclipse ends around 10 hours later, and the star's previous brilliance returns.

There are stars which actually do vary in intrinsic brilliance. Cepheid variables are yellow supergiants that rhythmically expand and contract, accompanied by a change in brightness. Delta Cephei, from which the class takes its name, varies between magnitudes 3.5 and 4.4 in a period of 5 days 9 hours. Other Cepheids have periods that vary between 2 to 50 days—the brighter the star, the longer its period. Cepheids are valuable in determining the size of the Milky Way and the nearness of other galaxies, because the length of their cycle is directly related to their brilliance. By comparing a Cepheid's apparent and actual brightness, its distance can be accurately gauged.

Long-period variables behave less dependably. Like Cepheids, long period variables are slowly pulsating, expanding, and contracting to produce changes in luminosity, but their cycles only follow a general pattern. Omicron Ceti (Mira, the "wonderful") is a well-known long period variable, with a period of around 330 days and a range between magnitudes 3 and 9.

Semi-regular and irregular variables are erratic, with few predictable patterns in their periods or brightness. Occasionally, stargazers

▲ Hubble Space Telescope took this close-up view of the dense star fields of the Milky Way in Sagittarius.

With just a few exceptions, only the nearest stars within our galactic spiral arm are visible to the unaided eye. Stars further away blend into a distant haze, seen from our perspective as a misty band, since we are looking directly through the plane of the galaxy where most of its stars and nebulae lie. In places, the Milky Way is bright enough to be discerned with the unaided eye from darker suburban locations. Seen from a really dark country site, far from light pollution, you can trace the Milky Way from horizon to horizon, and can discern considerable structure along it. Binoculars will reveal the individual stars within the Milky Way, in addition to many star clusters and galactic nebulae.

Stargazers in northern temperate latitudes can trace the Milky Way from Scorpius, through Sagittarius, Cygnus, Cassiopeia, Auriga, Gemini, and Canis Major. The galactic north pole is located in Coma Borenices, 90° above the plane of the Milky Way. Since the plane of the Milky Way is angled to the celestial equator, its height above the horizon depends on the time of the year and time of night. On dark Fall nights, the Milky Way tract running through the Cygnus region arches high overhead. An obvious dark rift running through Cygnus is a dense lane of galactic dust and gas, silhouetted against the bright star clouds beyond.

Open star clusters

Stars born within the same localized galactic cloud of dust and gas tend to cluster around a common center of gravity. However, over billions of years, the stars become spread out by wider gravitational forces.

▼ The Pleiades open star cluster, 100 million years old and 380 light years away.

Pleiades

Hundreds of young blue stars are contained within the Pleiades, the sky's brightest open star cluster. Five or six Pleiads can be seen with the unaided eye, but binoculars will reveal dozens of cluster members. Nearby, making up the head of Taurus the Bull, lie scattered the stars of the Hyades, the nearest open cluster to us, around 150 light years away. Aldebaran, a bright red star marking the bull's eye, is a foreground star and is not part of the cluster.

Praesepe

At the same time that the Pleiades and Hyades are high in the sky, another open star cluster, Praesepe, is also visible with the naked eye. Its member stars are slightly too faint to be individually resolved, so the cluster appears as a small misty patch. Through binoculars or a telescope it is obvious why the cluster is also known as "the Beehive"—around 40 stars ranging between 6th and 9th magnitude can be made out in a relatively small area of sky, a little larger than the size of the full Moon.

Between the Pleiades and Praesepe, just to the north, is a line of four bright and very distant telescopic open star clusters that run along the spine of the Milky Way. They include the bright glittering diamond brooch of M35 in Gemini, just visible with the unaided eye, and the fainter but equally fabulous clusters M36, M37, and M38 in Auriga. All of these are wonderful sights through a telescope at medium to high magnifications.

▾ Praesepe is ten times older than the ¹ades and lies around twice as far away ¹s.

Perhaps the most beautiful open cluster of them all lies in the Southern Cross. Called the "Jewel Box," the cluster delights telescopic observers with a display of around 50 stars, the brightest of which shine with a variety of colors.

Globular star clusters

Globular clusters are hefty star swarms, held together in a spherical mass by their mutual gravity. An average globular contains hundreds of thousands of old red stars and measures more than 100 light years in diameter. A halo of more than 150 globulars encircles the galaxy. They are extremely ancient, their stars having been among the first formed in the galaxy.

Binoculars and small telescopes show globulars as misty patches with bright centers, but individual stars within brighter globulars can be resolved through telescopes of 150mm or larger.

Located in Hercules and just about visible with the unaided eye, M13 is the brightest globular in northern skies. Centaurus hosts an even brighter globular, Omega Centauri. It is easily visible with the unaided eye as a fuzzy Moon-sized patch.

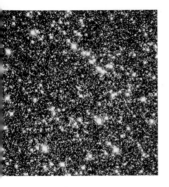

▲ A very close-up view of the ten million star-strong Omega Centauri globular cluster, the biggest in the galaxy.

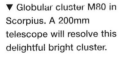

▼ Globular cluster M80 in Scorpius. A 200mm telescope will resolve this delightful bright cluster.

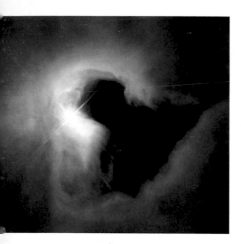

▲ NGC 1999 in Orion, a
bright reflection nebula
superimposed upon by a
keyhole-shaped dark
nebula.

▼ The Horsehead Nebula,
a dark column of interstellar
gas and dust in Orion.

Interstellar clouds

In addition to billions of stars, the disk of the
Milky Way contains enormous quantities of gas
and dust. Some of these interstellar clouds are
dense enough to obscure large tracts of the
galaxy, and they appear silhouetted against the
brighter background. Dark interstellar clouds
visible with the naked eye along the Milky Way
are prominent in the Cygnus and Sagittarius
regions, and there is a striking dark knot in the
southern constellation of Crux, a pitch-black
blob nicknamed the "Coalsack."

Most dark nebulae visible through telescopes
are physically associated with surrounding bright
nebulae; they can be seen simply because they
are blocking the light from a portion of the
nebula beyond. For example, a prominent dark
nebular wedge produces the well-known
"Shark's Mouth" in the Orion Nebula. Of all the
dark nebulae, the Horsehead Nebula is the
most famous. It too lies in Orion, although you
need a large telescope to catch a glimpse of it.
Set against the ruddy glow of hydrogen gas, this

dark pillar takes the form of a stylized chess knight, making one of astronomy's better-known simulacra.

Emerging from their galactic birthplace, the light from bright young stars illuminates the surrounding dust, producing a reflection nebula. Its brightness depends upon the size and density of the reflecting grains, and the color, brightness, and proximity of the nearby stars illuminating them. Reflection nebulae often have a pronounced blue color, caused by the reflective properties of carbon dust gains. Given dark skies, a hint of the reflection nebula surrounding the Pleiades open star cluster can be seen through large binoculars, particularly in the region of Merope, one of its brighter stars.

Intense ultraviolet light emitted by stars can make the hydrogen gas (the most abundant element in the universe) within their interstellar cocoons glow a red color. The colors of these emission nebulae can also be supplemented with green hues, caused by glowing oxygen gas, mixed with blue light produced by sulphur.

At the heart of the Orion Nebula glows a bright emission nebula whose subtle hues can be discerned with optical aid. It is part of a far larger but

much fainter diffuse nebula which spans virtually the whole of the constellation of Orion. Another beautiful combination of dark, reflection, and emission nebulae is the Trifid Nebula in Sagittarius, a deep-sky delight which takes its name from three dark lanes superimposed upon it.

▲ The Orion Nebula, the sky's brightest emission nebula.

▶ NGC 281, an emission nebula in Cassiopeia.

Planetary nebulae—stellar smoke rings

As we have seen, stars don't burn like ordinary matter, and their temperature and output of energy varies during the course of their lifetimes, as does their size. Toward the end of their lives, stars of the Sun's size swell up into red giants. But this phase of a star's life doesn't last long. Unable to maintain its hyperinflated state, the star's core contracts and shells of gas are puffed off into space. These shells become visible as planetary nebulae, so named because several bright examples resemble the shape of planets.

Shining by emission, a mix of hot gases within a planetary nebula can have the distinctive red hues of glowing hydrogen and the green of oxygen. Expanding from their parent star at velocities up to 20 miles per second, the gases eventually cool and fade from view. With an average lifetime of around 50,000 years—just 100,000th the average lifespan of its parent star—planetary nebulae are among the shortest-lived structures within the galaxy.

There are estimated to be around a thousand planetary nebulae in our galactic neck of the woods, but only a handful of them are easily visible through a small telescope. Easily the largest and brightest planetary nebula, the

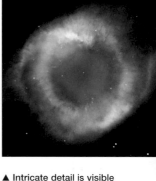

▲ Intricate detail is visible in this Hubble Space Telescope image of the Helix Nebula, a planetary nebula in Aquarius.

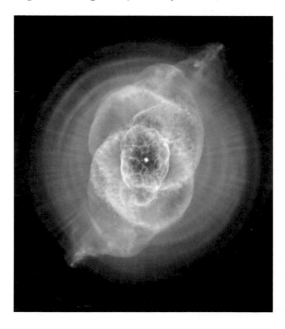

◄ The Cat's Eye Nebula in Draco displays detailed radial and concentric structure, as well as a series of bright, elliptical, inner shells.

Dumbbell Nebula in Vulpecula, has two brightly glowing lobes, giving it the telescopic appearance of an apple core.

▲ At the heart of the Crab Nebula lies a rapidly spinning pulsar, seen just above center left.

Planetary nebulae are among the most visually beautiful objects in the universe and vary enormously in appearance—no two are exactly alike. Some, such as the Ring Nebula in Lyra, have pronounced doughnut shapes. Others, like the Spirograph Nebula in Lepus, have a really intricate, lattice-like structure. A few actually do look like planets, notably the Saturn Nebula in Aquarius (complete with edge-on rings), and the Ghost of Jupiter Nebula in Hydra.

Supernova remnants

At the end of their lives, really big stars self-destruct in a supernova explosion, an event that produces so much light that it can briefly outshine its entire home galaxy. Only a few supernovae have ever been seen with the unaided eye, most famously the Crab Nebula supernova in 1054 AD. A 100mm telescope will show the Crab Nebula as a faint, featureless blob, and it requires a sizeable instrument to reveal any traces of the fine filaments that extend from the nebula's center. Still, it's the brightest supernova remnant visible. Like planetary nebulae, supernova remnants fade slowly over time as the expanding shell of material cools.

▶ Galaxies

A century ago, astronomers thought that the entire visible universe was contained within the boundaries of the Milky Way. We now know that our Milky Way is just one of billions of galaxies—each containing billions of stars—spread throughout the universe in space and time.

MUST KNOW

Big Bang fallout
Intense gravitational forces created supermassive black holes at the cores of infant galaxies.

▼ The beautiful spiral form of the Whirlpool Galaxy in Canes Venatici.

Galaxy types

Around dense galactic cores, the first stars formed from mainly hydrogen and helium gas. Density waves spreading throughout the galaxy like ripples in a pond prompted star-forming regions in outlying areas as dust and gas were compressed. The combination of new star formation with the rotation of the stars around the galactic center produces the wonderful '"grand design" pattern of spiral galaxies. Some spirals have tightly wound arms, while others are distinctly S-shaped, with loose arms. Barred spiral galaxies have elongated nuclear regions with arms spinning off either end.

star charts &

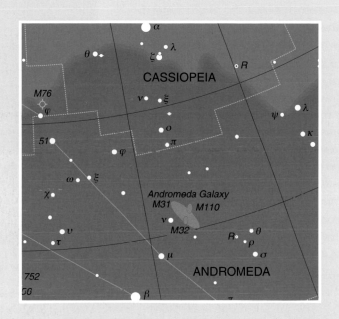

▶ NGC 4139, a barred spiral galaxy, appears in the same field of view as the quasar Mrk 205, an object that lies 14 times further away.

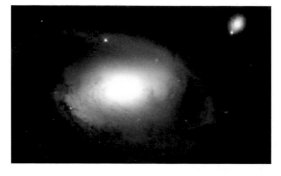

Elliptical galaxies are football-shaped and noticeably devoid of the dark lanes of dust and gas commonly seen in spirals. Irregular galaxies have little geometric form. Many have been gravitationally disrupted by a nearby galaxy or are actively involved in a direct collision with another galaxy. The resulting shapes created in such titanic encounters can be spectacular.

Intergalactic views

Anyone with average eyesight and dark skies can see the Andromeda Galaxy with the naked eye alone (see page 110). So far away that its light takes 2.5 million years to reach us, the galaxy appears as an elongated smudge as wide as the full Moon through binoculars and small telescopes. Structure can be seen through larger telescopes, including dark lanes and nebulae in its spiral arms. Galaxies M32 and M110, the dwarf satellites of the Andromeda Galaxy, can also be found in the same low-power field of view.

Binoculars and small telescopes will reveal dozens of more distant galaxies, including the Whirlpool Galaxy in Canes Venatici and the duo of the Cigar Galaxy and Bode's Galaxy in Ursa Major, which fit into the same low-power telescopic field.

◀ These two spiral galaxies in Canis Minor (NGC 2207 and IC 2163) are colliding with each other.

DEEP SPACE

97

showpieces

The stargazer needs only a good star map and a keen pair of eyes to learn their way around the skies. After a while, the mysterious constellations and their strange-sounding stars become well-known friends, though the romance of the skies never fades.

▶ # Constellations

In ancient times, patterns of stars represented the outlines of sacred objects, mythical beings, and supernatural creatures that featured in the legends of various cultures around the globe.

A catalogue compiled by the Greek astronomer Ptolemy in 150AD featured 48 constellations, including the 12 constellations of the zodiac. Many of these constellations had been in use for thousands of years previously.

Constellations are a handy means of identifying and locating objects in the sky. Eighty-eight constellations of various shapes and sizes are officially recognized today, all contained within neatly defined boundaries, making the celestial sphere look like an enormous 3D jigsaw puzzle. In addition to Ptolemy's original 48, more recent additional constellations include those covering the southern skies.

Stars

Each constellation contains an assortment of stars of varying brightness, and the Greek alphabet is used to designate the brightest of these stars. For example, the brightest star in Taurus is Alpha Tauri, the second brightest is Beta Tauri, then Gamma Tauri, and so on. A stargazer will find it is worth becoming familiar with the Greek alphabet—at least, the names and symbols for the first dozen or so letters, which are commonly featured on most star maps.

Astronomers have names for individual bright stars, too. Alpha Tauri also goes by the name of Aldebaran—an ancient Arabic name, meaning the "follower." Star names derived from Arabic sources feature prominently, and there are also a number with ancient Greek names, such as

MUST KNOW

Greek alphabet
This shows the name of the letter, its equivalent in the Roman alphabet, and the Greek symbol.

	Rom	Gk
Alpha	a	α
Beta	b	β
Gamma	g	γ
Delta	d	δ
Epsilon	e	ε
Zeta	z	ζ
Eta	h	η
Theta	th	ϑ
Iota	i	ι
Kappa	k	κ
Lambda	l	λ
Mu	m	μ
Nu	n	ν
Xi	x	ξ
Omicron	o	o
Pi	p	π
Rho	r	ρ
Sigma	s	σ
Tau	t	τ
Upsilon	u	υ
Phi	ph	φ
Chi	ch	χ
Psi	ps	ψ
Omega	o	ω

Sirius (meaning "searing hot"). A grand mix of cultures through the ages has created the nomenclature of the heavens used today.

The star charts

As can be seen from the index charts featured overleaf, the 20 star charts featured in this section give a complete coverage of the entire celestial sphere. Each chart overlaps generously with its neighboring charts, enabling them to be referred to more easily.

Chart 1 includes the northern celestial polar regions down to a declination of around 65° N. Charts 2 to 7 cover a region from above 70° to below 20° N declination. Charts 8 to 13 cover the celestial equatorial regions in a band from around 25° N to 25° S. Charts 14 to 19 cover a region from above 20° S to below 70° S declination. Chart 20 includes the southern celestial polar regions to a declination of around 65° S.

▼ The two red lines represent the limit of visibility of the celestial sphere from Vancouver (upper line), and Cape Canaveral (lower line).

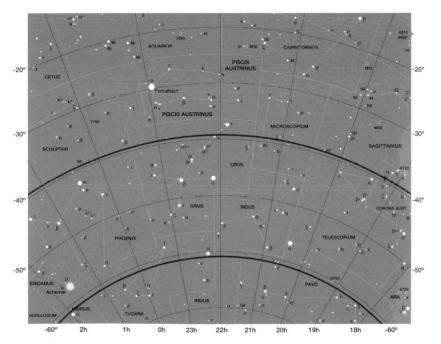

All 88 constellations are depicted on the charts. Stars down to magnitude 5.5 are drawn, which is around the limit of visibility with the unaided eye from a reasonably dark site. Accompanying each chart is a description of each constellation, special attention being paid to those objects visible through binoculars and small telescopes.

The sky's the limit

To find the limit of your view of the skies below the celestial equator, simply deduct the figure for your latitude on Earth from 90°. Note that atmospheric effects and pollution cause an effect known as extinction, where stars become obscured near the horizon; only the brightest stars can be seen within a few degrees of the horizon.

INDEX—NORTHERN HEMISPHERE

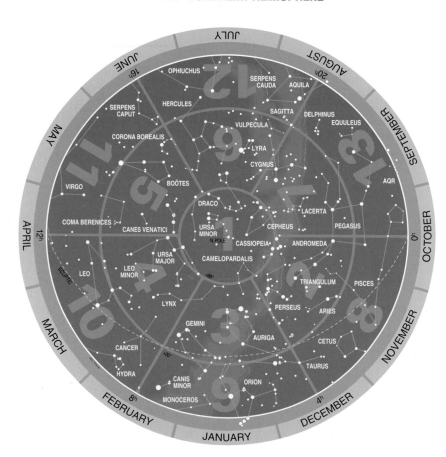

Stargazers in Columbus, Ohio, say at latitude 40° N, are able to view stars as far south as 50° S of the celestial equator. This is sufficiently far south to view the whole of Piscis Austrinus and Pyxis, and theoretically the far northern part of Grus. From Cape Canaveral in Florida, USA, at latitude 29° N, stargazers are treated to a view of stars down to a declination of 61° S; from here, only a small region around the southern celestial pole is denied the stargazer. The same calculation can be made for stargazers in the southern hemisphere. From Melbourne, Australia, at 38° S, stars to 52° N may be viewed. Tourists from the northern hemisphere may find it incredible that far northern constellations such as Perseus, Auriga, and Canes Venatici can be seen without difficulty from southern Australia.

INDEX—SOUTHERN HEMISPHERE

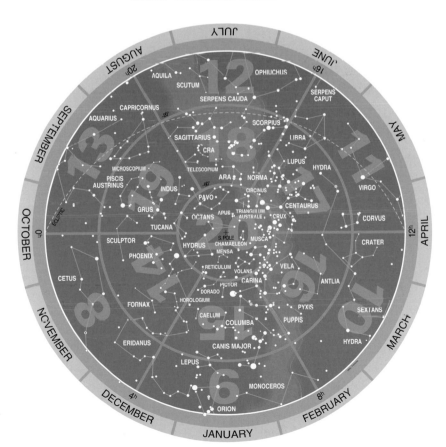

▶ # Chart 1

Home to the north celestial pole, this region is always on view from the USA, Canada, Europe, and the UK. The chart includes all of the circumpolar constellations of Cepheus and Ursa Minor, along with much of Camelopardalis, Cassiopeia, and Ursa Major. The Milky Way passes through the charts, from northern Cygnus, through Cepheus, Cassiopeia, and into northern Perseus.

Ursa Minor (the Little Bear)

UMi/Ursae Minoris; highest at midnight: early January

Sometimes referred to as the Little Dipper, Ursa Minor is a small but significant constellation, since it encloses the north celestial pole. Its brightest star, Polaris (magnitude 2), is located within just 1° of the pole, and is referred to as the Pole Star. A phenomenon called "precession," caused by the slow movement of the Earth's axis, will bring Polaris within half a degree of the pole in around 100 years' time. Polaris is useful to aim at when aligning an equatorial telescope for a quick observing session. A telescope roughly aligned with Polaris will keep an object within the field of view of a medium power eyepiece for a long time before requiring adjustment. A telescope trained on Polaris will reveal it to be a double star, with an 8th magnitude companion.

Beta UMi (Kochab) and Gamma UMi (Pherkad), the end stars of the Little Dipper, are sometimes called the Guardians of the Pole. A keen naked eye will discern a faint star close to Pherkad (about half the Moon's diameter away)—this is an unrelated foreground star.

MUST KNOW
Not so bright Ursa Minor does not contain any bright deep sky objects.

▶ Open star cluster IC1396 in Cepheus.

Cepheus (King of Ethiopia)

Cep / Cephei Highest at midnight: Late February

Cepheus contains several reasonably bright stars and can easily be located because of its shape, which resembles a house with a steep pointed roof.

Among the stellar delights of Cepheus is Beta Cep, a double star of magnitudes 3.2 and 7.9 that can be resolved through a small telescope. Delta Cep has a blue mag 6.3 companion, and the pair is easy to separate through a small telescope.

Mu Cep is a red supergiant whose striking color has earned it the nickname of the Garnet Star. Binoculars will show its ruddy hue to good effect. Mu Cep is one of the biggest stars visible with the unaided eye—if placed in the position of the Sun, the surface of its enormous bloated sphere would extend almost out to the orbit of Saturn.

Xi Cep is a double star comprising a magnitude 4.4 blue star and a magnitude 6.5 orange giant, easily resolvable through a small telescope.

An extension of the Milky Way nudges into the southern part of Cepheus, and a couple of lovely open star clusters can be found in this vicinity—IC 1396 and NGC 7160—both delightful to view through a 150mm telescope. IC 1396 is embedded within a sizeable patch of nebulosity; it is also known as the Elephant's Trunk Nebula, because of a prominent dark sinuous dust lane which is visible on photographs. Large binoculars will reveal IC1396 as a misty patch. NGC 7160 is a small, compact star cluster; around 30 of its stars are visible through a 200mm telescope, half a dozen of the brighter ones standing out from the rest.

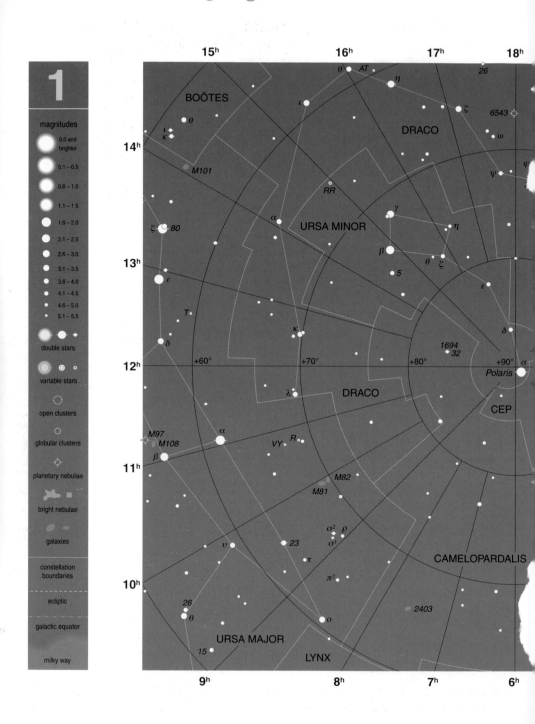

1

magnitudes

- 0.0 and brighter
- 0.1 – 0.5
- 0.6 – 1.0
- 1.1 – 1.5
- 1.6 – 2.0
- 2.1 – 2.5
- 2.6 – 3.0
- 3.1 – 3.5
- 3.6 – 4.0
- 4.1 – 4.5
- 4.6 – 5.0
- 5.1 – 5.5

double stars

variable stars

open clusters

globular clusters

planetary nebulae

bright nebulae

galaxies

constellation boundaries

ecliptic

galactic equator

milky way

15ʰ   16ʰ   17ʰ   18ʰ

BOÖTES

θ   AT
η
ι
DRACO
26
θ
ι
κ
14ʰ
ω
M101
6543
ψ
φ
RR
α
URSA MINOR
γ
η
ζ   80
β
ζ
13ʰ
θ   ζ
ε
5
ε
T
δ
κ
1694   δ
32
12ʰ   +60°   +70°   +80°   +90°   α
Polaris
λ
DRACO
CEP
M97
α
M108
R
β
VY
11ʰ
M82
M81
σ²   ρ
σ¹
υ   23
CAMELOPARDALIS
τ
π²
10ʰ
2403
26
θ
o
URSA MAJOR
15
LYNX
9ʰ   8ʰ   7ʰ   6ʰ

NGC 752 is a large open cluster made up of more than 60 faint stars spread over an area larger than the full Moon. The cluster is visible as a misty patch through binoculars, and its individual stars are resolvable through a 100mm telescope.

The Blue Snowball (NGC 7662) is a bright, 9th magnitude planetary nebula, visible through small telescopes as a fuzzy blue spot.

▲ The Bubble Nebula (NGC 7635) in Cassiopeia.

Cassiopeia (Queen of Ethiopia)
Cas/Cassiopeiae; highest at midnight: early October

The prominent W-shaped asterism of five bright stars makes Cassiopeia one of the easiest constellations to recognize. Eta Cas is a nice double star with a mag 3.5 yellow primary and a red mag 7.5 companion, easily visible through a small telescope.

Cassiopeia is a joy to scan with binoculars, as a bright section of the Milky Way flows across the constellation, engulfing the "W." A treasure trove of bright open clusters lies within its boundaries. Containing around 30 stars, the bright compact cluster of M103 is best seen at higher magnifications. The Owl Cluster (NGC 457) is a loose assembly of around 100 fairly bright stars arranged in distinct lines; its two brightest stars shine like an owl's eyes. NGC 663 is a beautiful binocular cluster containing around 80 stars. On the far western side of Cassiopeia, the Scorpion Cluster (M52) contains around 100 stars, the brightest of which form a splendid S-shape.

Triangulum (the Triangle)
Tri/Trianguli; highest at midnight: late October

Triangulum is one of the sky's smallest and least prominent constellations, made up of three faint stars that form an elongated triangle. Insignificant though Triangulum appears, it hosts one of the nearest galaxies—the Pinwheel Galaxy (M33)—a face-on spiral 2.7 million light years distant, visible with the naked eye from a dark sky site. It has a low surface brightness, so although binoculars may easily show it, it can be missed using a telescope with a higher magnification.

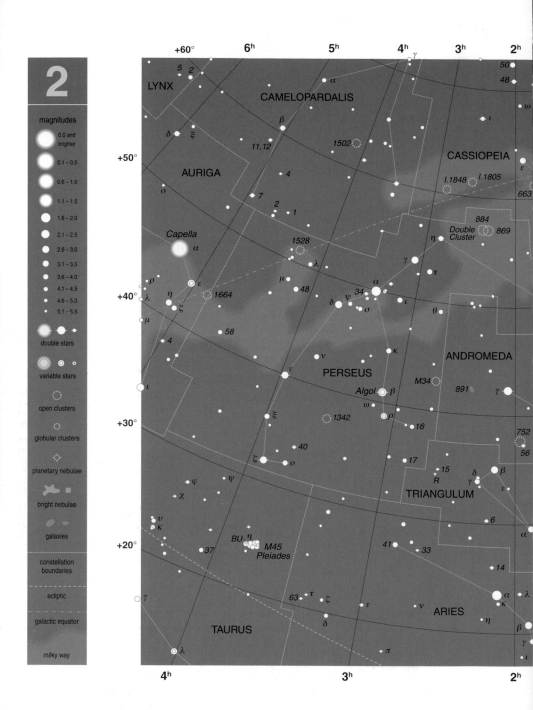

2

magnitudes

- 0.0 and brighter
- 0.1 – 0.5
- 0.6 – 1.0
- 1.1 – 1.5
- 1.6 – 2.0
- 2.1 – 2.5
- 2.6 – 3.0
- 3.1 – 3.5
- 3.6 – 4.0
- 4.1 – 4.5
- 4.6 – 5.0
- 5.1 – 5.5

double stars

variable stars

open clusters

globular clusters

planetary nebulae

bright nebulae

galaxies

constellation boundaries

ecliptic

galactic equator

milky way

LYNX

CAMELOPARDALIS

5 2

α

β

δ ξ

11,12

1502

CASSIOPEIA

ι

ω

ε

I.1848   I.1805

663

AURIGA

4

o

7

2   1

1528

884

Double Cluster   869

η

Capella

α

λ

μ

48

34

α

γ

τ

ρ

ε

1664

δ   ψ

σ

ι

θ

η   η

ξ

ν

κ

ANDROMEDA

μ

58

ε

PERSEUS

M34

891

γ

4

Algol   β

ι

ω

752

ο

16

56

40

1342

17

15

R   γ   δ   β

ζ   ο

ε

TRIANGULUM

φ   ψ

6

χ

α

ν   κ

BU   η

41   33

14

37

M45
Pleiades

α   λ

κ

63   τ   ξ

ε

ν   η

ARIES

β

δ

γ

γ

π

ι

TAURUS

λ

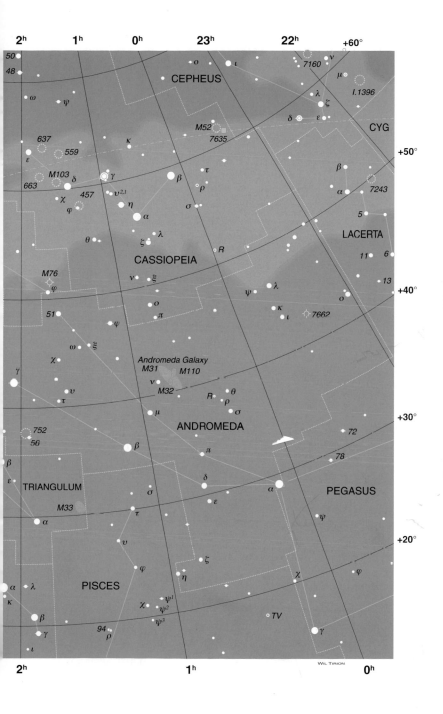

▶ Chart 3

The bold constellation of Auriga, bordered in the west by the bright Perseus and in the east by the much fainter Lynx, dominates this section of the northern skies.

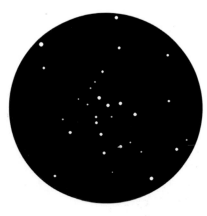

▲ Star cluster M36 in Auriga, observed through a small telescope.

Auriga (the Charioteer)
Aur/Aurigae; highest at midnight: mid-December

Broad, bright, and easily recognizable, Auriga rides high overhead on winter nights in the US. Southwest of brilliant Alpha Aur (Capella) lies the small triangular asterism of the Kids.

Three of the brightest clusters in Auriga—M36, M37, and M38—are easily visible through binoculars as misty patches. Of these, M37 is the biggest and brightest; an orange 9th magnitude star lies at the center of around 150 fainter stars, resolvable with a 150mm telescope. M38 is made up of around 100 stars below tenth magnitude, the brightest of which form a startling cross-shape.

Cameleopardalis (the Giraffe)
Cam/Cameleopardalis; highest at midnight: mid-December

Despite being a large circumpolar constellation, Cameleopardalis is one of the least recognized northern constellations. This isn't surprising, since its brightest star, Beta Cam, shines a feeble mag 4. The Milky Way nudges into western Cameleopardalis, and here NGC 1502, an open cluster of around a dozen stars, is resolvable through a small telescope. To its immediate west, and in the same low power field, a lovely line of around a dozen bright stars called Kemble's Cascade can be traced across more than 2° of sky.

Gemini (the Twins)
Gem/Geminorum; highest at midnight: early January

A sizeable, easily recognized constellation, Gemini is a bright and well-defined constellation lying between Auriga and Canis Minor. Its two main stars, Alpha (Castor) and Beta Gem (Pollux), are instantly identifiable.

Pollux, slightly the brighter of the pair, has a decidedly orange hue. Castor is a famous multiple star, with its two brightest components (mags 1.9 and 3) resolvable through a good 60mm telescope.

The western part of Gemini is immersed in the Milky Way. Here, the open cluster M35 can be glimpsed with the unaided eye, a couple of degrees northwest of Eta Gem. Binoculars show it well. In an area the size of the full Moon, M35 contains about 80 stars, some of which are arranged into a prominent curved chain.

A couple of degrees southeast of Delta Gem can be found the Eskimo Nebula (NGC 2392), a bright 8th magnitude planetary nebula that shows up as a greenish blob at low magnifications.

Perseus (hero of Greek mythology)
Per/Persei; highest at midnight: mid-November

Crossed in the north by the Milky Way, magnificent Perseus contains a number of bright stars and open clusters. Near Alpha Per (Mirfak) lies Melotte 20, a large, loose star cluster made up of a snaking chain of stars. Beta Per (Algol) is a famous eclipsing binary. Every 2.87 days it drops from mag 2.1 to 3.4, easily monitored with the unaided eye. The Spiral Cluster (M34) can just be discerned with the unaided eye 5° west of Algol.

Eta Per is a nicely colored double star, resolvable through a small telescope, with an orange mag 3.8 primary and a blue mag 8.5 companion.

In the northwestern corner of Perseus, the Double Cluster (NGC 869 and NGC 884) is one of the most breathtaking sights in the heavens. These two bright open clusters—each the diameter of the full Moon—lie side by side, and can be glimpsed as a hazy patch with the unaided eye. A low power view accommodates both clusters, revealing hundreds of stars. Several red stars can be discerned near NGC 884, contrasting nicely with the cluster's profusion of blue stars.

At magnitude 10, the Little Dumbbell Nebula (M76) is the faintest Messier object. Resembling an apple core, this dim planetary nebula can be seen through a 150mm telescope.

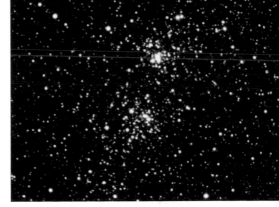

▶ The magnificent Double Cluster in Perseus.

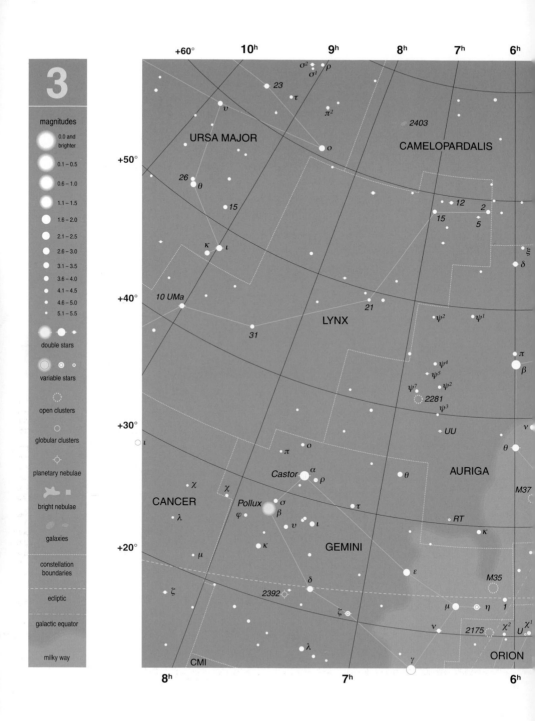

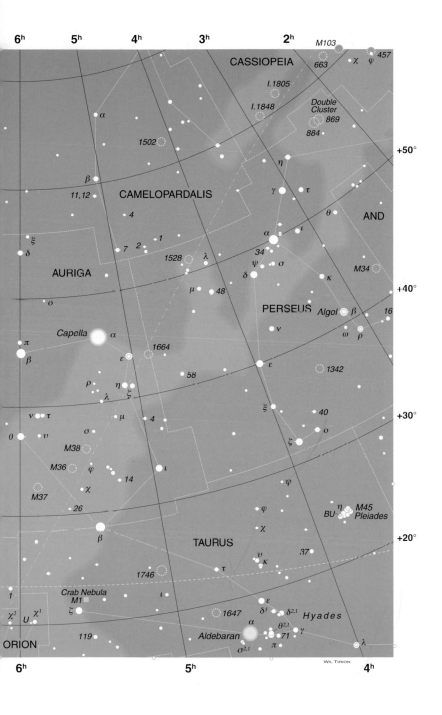

6ʰ 5ʰ 4ʰ 3ʰ 2ʰ

CASSIOPEIA

M103

663 χ φ 457

I.1805

I.1848

Double Cluster
869
884

1502

η

+50°

α

β

CAMELOPARDALIS

γ τ

θ

AND

11,12

4

α ι

ξ

7 2 1

34

M34

δ

1528 λ

ψ σ

κ

AURIGA

μ 48

δ

PERSEUS Algol β 16

o

Capella α

1664

ν

ω ρ

π

ε

58

ε

1342

β

ρ η ζ

λ ζ

ξ

40

+40°

ν τ

μ 4

o

+30°

θ v

σ

ζ

M38

M36 φ ι

14

ψ

M37

χ

φ

η M45
BU Pleiades

26

χ

+20°

β

TAURUS

υ 37
κ

1746

τ

Crab Nebula
M1

ι

ε

1 δ³ δ²¹ Hyades

χ² χ¹

ζ

1647 θ²¹ 71 γ

U

α

ORION

119

Aldebaran

π λ

σ²¹

6ʰ 5ʰ 4ʰ

WIL TIRION

▶ # Chart 4

Ursa Major, one of the sky's most easily recognizable constellations, sprawls across much of this section of the northern sky. The Milky Way is absent from these charts, since we are peering high above the galactic plane.

Leo Minor (the Little Lion)
LMi/Leo Minoris; highest at midnight: late February

Consisting of a smallish grouping of relatively faint stars located between Ursa Major and Leo, stargazers often overlook Leo Minor. The lack of any bright deep sky objects makes the constellation somewhat neglected, although large amateur telescopes will reveal a number of faint galaxies.

Lynx
Lyn/Lyncis; highest at midnight: late January

Lynx is made up of a collection of dim stars between Ursa Major and Auriga. 3rd magnitude Alpha Lyn is the brightest of its stars, and likely to be the only one that can be made out with the unaided eye from a light polluted urban location.

12 Lyn is a triple star, with close blue primary components of mag 5.5 and 6.1, just separable with an 80mm telescope, and a fainter but more distant mag 7 component that can easily be resolved with a smaller instrument. 19 Lyn is a more widely separated triple star, though each component (at mag 5.8, 6.8 and 7.6) is easily visible when viewed through a 60mm telescope.

The Intergalactic Wanderer (NGC 2419) is a small, exceedingly distant globular cluster, easy to see through a 150mm telescope, although it requires a much larger instrument to resolve into stars.

Ursa Major (the Great Bear)
UMa/Ursae Majoris; highest at midnight: early March

Seven of the brightest stars within Ursa Major make up an asterism variously called the Big Dipper or the Plough. While this asterism itself doesn't much look like a bear, a little time spent in tracing the traditional outline comprising the remainder of the constellation's bright stars will convince any stargazer that the ancients who named it had an extremely good eye for form.

The two front stars of the Big Dipper, Alpha UMa (Dubhe) and Beta UMa (Merak) are known as the Pointers, since an imaginary line extending from them leads to Polaris and the north celestial pole. Zeta UMa (Mizar), the second star of the Dipper's handle, has a fainter mag 4 partner,

▲ The Big Dipper (or Plough)—a star pattern familiar to all stargazers in the northern hemisphere.

80 UMa (Alcor), which is visible with the unaided eye. Mizar itself is a close double star, with components of mag 2.2 and 3.8, separable even with a small telescope.

A pair of galaxies bright enough to be seen through binoculars, Bode's Galaxy (M81) and the Cigar Galaxy (M82), lie in the far north. Just half a degree apart, the pair is visible in the same low-power telescopic field. While M82 is almost edge-on to us, M81 is tilted at less of an angle. On the other side of the constellation, the face-on spiral galaxy M101 is visible through binoculars as a circular smudge, and appears mottled through a 200mm telescope.

The Owl Nebula (M97), a faint planetary nebula, appears as a pale disk about twice the diameter of Jupiter through a 150mm telescope. The dark eyes of the owl, so obvious in many images, are rather elusive and require at least a 250mm telescope to discern. M108, a bright, sizeable and nearly edge-on galaxy, can fit into the same low-power telescopic field as M97. Bright condensations within M108 can be discerned through a 150mm telescope.

WATCH OUT!

Galaxy M109 location
Although M109 can be identified fairly easily, being a little more than half a degree east of Gamma UMa, its low surface brightness makes it more of a challenge to observe.

▼ The Owl Nebula (M97), a delightful planetary nebula in Ursa Major.

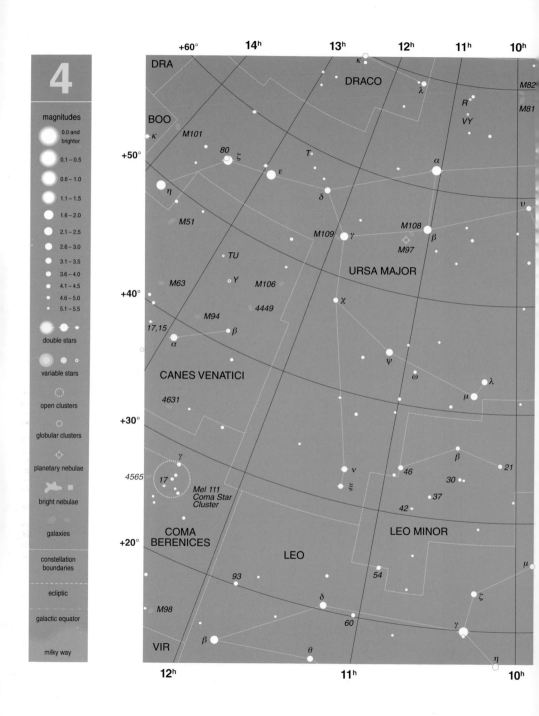

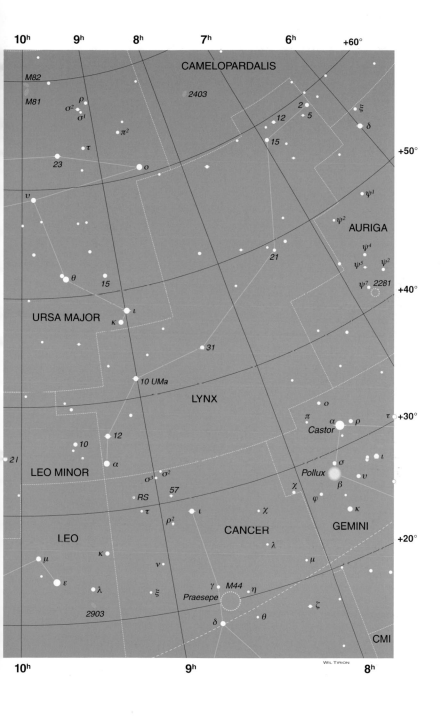

CAMELOPARDALIS

M82

M81

2403

σ² ρ

σ¹

π²

τ

23

ο

υ

θ   15

ι

κ

URSA MAJOR

+50°

+60°

2

5

12

15

ξ

δ

ψ¹

ψ²   AURIGA

ψ⁴

ψ⁵   ψ²

ψ⁷   2281   +40°

21

31

10 UMa

LYNX

12

10

α

21

o

π   α   ρ   τ   +30°

Castor

σ   ι

Pollux   υ

χ   β

φ   κ

GEMINI

LEO MINOR

σ³   σ²

RS   57

τ

ρ²   ι   χ

CANCER

λ

LEO

κ

ν

μ

ε   λ

ξ

γ   M44   η

Praesepe

μ

ζ

2903

δ   θ

CMI

+20°

▶ # Chart 5

Our view towards Boötes and Canes Venatici remains well
north of the galactic plane, so the Milky Way is again absent.
There is still an abundance of stellar delights and far off
galaxies spread throughout the region.

Boötes (the Herdsman)
Boo/Boötis; highest at midnight: early May

The brightest star in Boötes, the orange giant Alpha Boo (Arcturus) can
be located by tracing a slightly curving line from the tail of Ursa Major.
Many stargazers trace out the shape of Boötes by imagining a large kite,
with Arcturus at its tail end and Beta Boo at its apex; the image is
completed by a trail of stars west of Arcturus, representing the kite's
trailing ribbon.

   Although Boötes is a large constellation, it contains no bright deep sky
objects. It makes up for this with a number of fine double stars. Epsilon
Boo is a lovely close double with an orange primary (mag 2.5) and blue
companion (mag 4.6), and requires a 100mm telescope at x100 to
resolve well. Doubles that are easily visible through a 60mm telescope
include Iota Boo, a wide double of mags 4.8 and 8.3; Kappa Boo, an
easy double of mag 4.5 and 6.6; Pi Boo, mag 4.5 and 5.8; and Xi Boo, a
beautiful yellow-orange double of mag 4.7 and 7.

Canes Venatici (the Hunting Dogs)
CVn/Canum Venaticorum; highest at midnight: early April

Canes Venatici nestles snugly beneath the tail
of Ursa Major, but only two of its stars—Alpha
CVn (Cor Caroli) and Beta CVn—are bright
enough to be seen from the average suburban
site with the naked eye. Cor Caroli is a nice
telescopic double of mags 2.9 and 5.6.

   Globular cluster M3 lies at the far southern
edge of Canes Venatici. It is large and bright,
easily visible as a misty patch through
binoculars; its outer stars can be resolved with
a 150mm telescope.

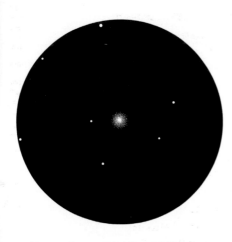

▲ Globular Cluster M3 in Canes Venatici, as
seen through a 150mm refractor.

The Whirlpool Galaxy (M51) is an easy binocular object, but a 200mm telescope is required to reveal a hint of the object's spiral

▲ Whirlpool Galaxy (M51) in Canes Venatici.

arms. M51's companion galaxy, NGC 5195, is visible through a 100mm telescope, and is joined to M51 by a faint bridge which can be discerned through a 300mm telescope.

Through a 200mm telescope, the Sunflower Galaxy (M63) appears as a featureless elliptical patch with a bright condensed nucleus. Another telescopically featureless galaxy with a condensed core, is the face-on Cat's Eye Galaxy (M94); it lies just north of the line connecting Cor Caroli with Beta CVn, so it is easy to locate. M106 is a large and bright spiral galaxy, and its elongated shape and point-like core discernible even through a small telescope.

Corona Borealis (the Northern Crown)

CrB/Coronae Borealis; highest at midnight: late May

A small but delightful constellation, the main pattern of Corona Borealis is made up of a semicircle of seven bright stars. Several doubles are easily separable even through small telescopes, including Zeta CrB (mag 5 and 6), Nu CrB (a wide 5th magnitude pair), and Sigma CrB (mag 5.6 and 6.6).

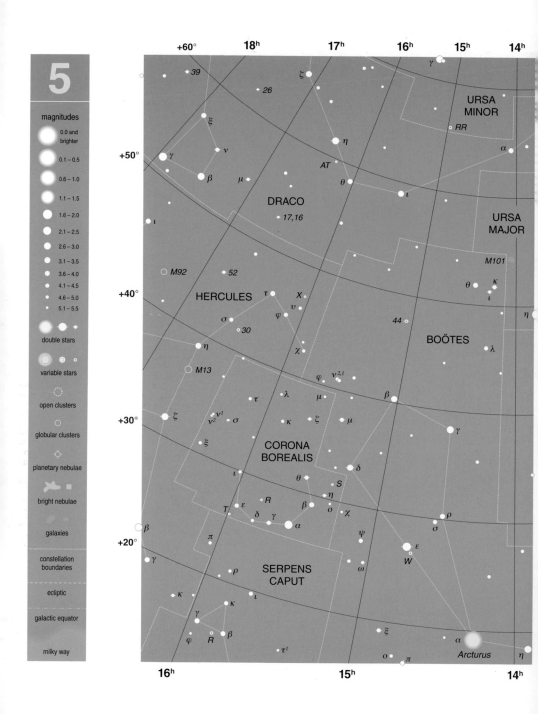

5

magnitudes

○ 0.0 and brighter
○ 0.1 – 0.5
○ 0.6 – 1.0
○ 1.1 – 1.5
○ 1.6 – 2.0
○ 2.1 – 2.5
○ 2.6 – 3.0
○ 3.1 – 3.5
· 3.6 – 4.0
· 4.1 – 4.5
· 4.6 – 5.0
· 5.1 – 5.5

double stars

variable stars

open clusters

globular clusters

planetary nebulae

bright nebulae

galaxies

constellation boundaries

ecliptic

galactic equator

milky way

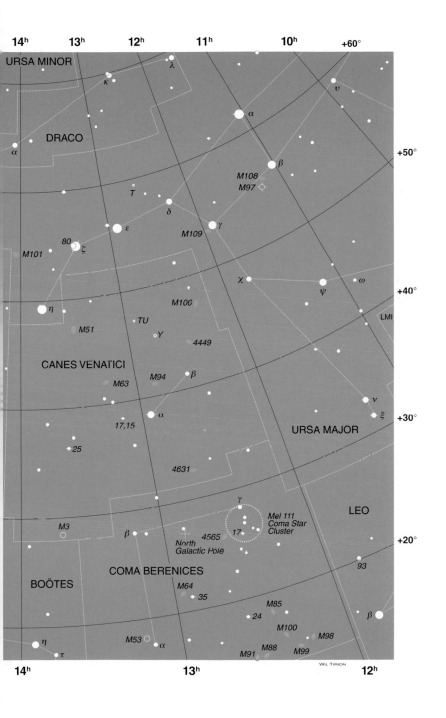

14ʰ   13ʰ   12ʰ   11ʰ   10ʰ   +60°

URSA MINOR

λ

κ

DRACO

α

+50°

α

M108

M97

τ

δ

β

ε

M109   γ

80   ζ

M101

χ   ψ   ω   +40°

η

LMI

M106

TU

M51   Y   4449

CANES VENATICI

M63   M94   β

ν

ξ   +30°

17,15   α

URSA MAJOR

25

4631

γ

LEO

M3   Mel 111

Coma Star

Cluster

β   4565   17

North   +20°

Galactic Pole

COMA BERENICES   93

BOÖTES   M64   β

35

M85

24

η   M100   M98

τ   M53   α   M88   M99   β

M91

Wil Tirion   12ʰ

14ʰ   13ʰ

▶

Chart 6

While stargazers in urban areas might struggle to glimpse the bright section of the Milky Way and the fainter stars depicted on the chart, there should be no trouble in finding Vega in Lyra and Deneb in Cygnus, two of the stars of the northern Summer Triangle asterism (the other star is Altair in Aquila—see chart 12).

Draco (the Dragon)
Dra/Draconis; highest at midnight: early July

Draco sprawls across a huge portion of the northern circumpolar region, covering an area of more than a thousand square degrees, from Camelopardalis to Cygnus. Despite being so large, Draco's has only a few reasonably bright stars, these ranging between the 2nd and 3rd magnitude.

Mu Dra is a close telescopic double star with white components of mag 4.9 and 5.6; the pair is slowly moving apart, and they can now be comfortably resolved with a 100mm telescope, and a good test for a 60mm telescope. Psi Dra is much easier to resolve; binoculars will reveal the yellow stellar duo of mags 4.6 and 5.8. Binoculars can split the two wide components of 39 Dra (mag 5 and 7.4); a telescope will show the mag 8 companion of the brighter star.

Hercules (hero of Greek mythology)
Her/Herculis; highest at midnight: early June

A large constellation, identifiable by the Keystone asterism made up of four of Hercules' brighter stars—Pi, Eta, Epsilon, and Zeta Her. These stars, combined with Beta and Delta Her further south, make up a familiar big butterfly-shaped asterism. The constellation stretches considerably further north, south, and east of this, making Hercules the sky's 5th largest constellation.

Alpha Her is a splendid red giant with a close green companion of mag 5.4. Gamma Her and 95 Her are also nice telescopic doubles.

The Great Globular Cluster (M13) is the northern sky's brightest example of its type. Lying just 2.5° south of Eta Her, M13 is easy to locate —indeed, it is faintly visible with the unaided eye. Binoculars show it to be an extensive fuzzy patch around half the apparent diameter of the full Moon. Viewed through a 150mm or larger telescope, the cluster is an amazing sight; the brightest of its 300,000 outlying stars can be resolved, and these appear to be arranged in several distinct radial lines. Hints of darker lanes can be discerned within the cluster's outer regions; photographs do not show these features well, but our perception through the eyepiece produces a different impression.

M92 is probably the northern sky's second most beautiful globular cluster. Located north of the Keystone, it receives less attention than its brighter sibling, but it is in many ways just as spectacular. Its outer stars can be resolved through a 150mm telescope, and the cluster is smaller and more compact and spherical than those of M13.

Lyra (the Lyre)
Lyr/Lyrae; highest at midnight: early July

The compact constellation of Lyra is one of the best known constellations of northern summer skies. Its western margin is clipped by a rich, wide section of the Milky Way.

Beta Lyr is a beautiful double star, the primary being a white variable star of mag 3.3-4.4, the companion a blue mag 7.2. The Double-Double Star (Epsilon Lyr) is a famous multiple. The main pair is wide enough to be separated with binoculars, while each of these is a close telescopic double of mags 4.6 and 5.3, and mags 4.7 and 6.1, each comfortably resolvable through a 100 mm telescope.

The Ring Nebula (M57) is one of the sky's best-known planetary nebulae. It can easily be found, as it is located almost directly between Beta Lyr and Gamma Lyr. M57 is rather small, and binoculars will show it as an almost starlike point of light. Through a telescope at a medium to high magnification it resembles a sharply defined glowing smoke ring.

The brightest of Lyra's other deep sky delights is the globular cluster M56, many of its stars resolvable through a 200mm and set in a lovely rich galactic star field.

▼ M57, the Ring Nebula in Lyra, a delightful planetary nebula.

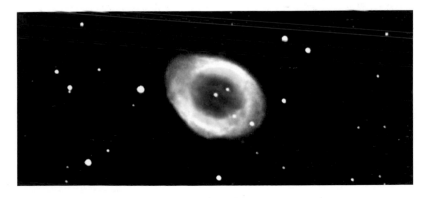

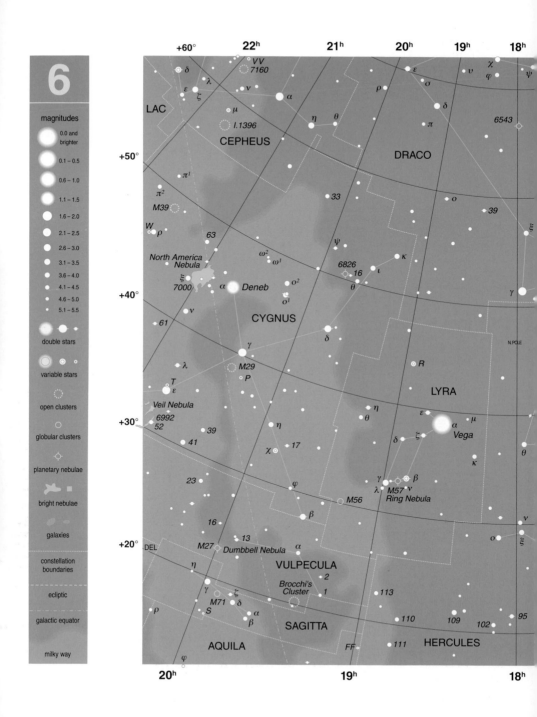

6

magnitudes

- 0.0 and brighter
- 0.1 – 0.5
- 0.6 – 1.0
- 1.1 – 1.5
- 1.6 – 2.0
- 2.1 – 2.5
- 2.6 – 3.0
- 3.1 – 3.5
- 3.6 – 4.0
- 4.1 – 4.5
- 4.6 – 5.0
- 5.1 – 5.5

double stars

variable stars

open clusters

globular clusters

planetary nebulae

bright nebulae

galaxies

constellation boundaries

ecliptic

galactic equator

milky way

+60°   22ʰ   21ʰ   20ʰ   19ʰ   18ʰ

+50°

+40°

+30°

+20°

20ʰ   19ʰ   18ʰ

LAC

CEPHEUS

DRACO

6543

VV
7160

δ
λ
ε ζ
ν
α
μ
I.1396

ε
υ
χ
φ
ψ
ρ
σ
δ
π

η θ

π¹
π²
M39

W ρ
63

33

39

ο
39

ξ

North America
Nebula
ξ
7000
α Deneb

ω² ω¹

ψ

6826
16 ι
θ

κ

γ

61
ν

CYGNUS

δ

N.POLE

λ
M29
P

T
ε
Veil Nebula
6992
52
39
41

η
θ

R

LYRA

η
ε
δ ξ
γ

α
Vega

μ

κ

23

χ 17

φ

γ
λ M57
Ring Nebula
β
ν

θ

16

β

M56

ν

ο

ξ

+20° DEL
M27
Dumbbell Nebula
13
α

VULPECULA

η

γ
ζ
M71 δ
S
ρ

Brocchi's
Cluster
2
1

α
β

SAGITTA

113

110

109

102

95

AQUILA
φ

FF

111

HERCULES

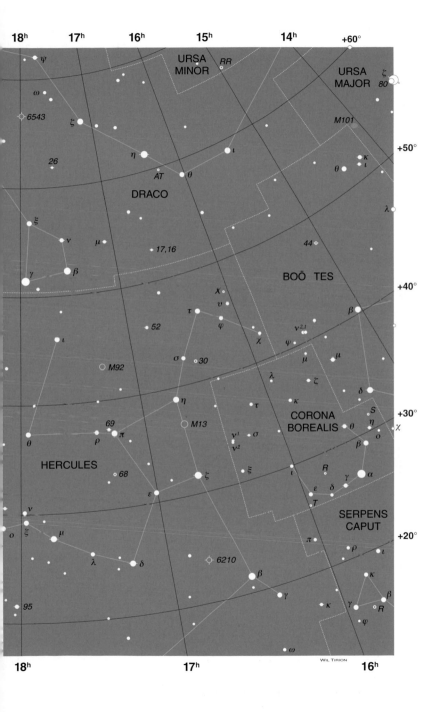

WIL TIRION

▶ # Chart 7

A bright section of the Milky Way slices through the sky from Cassiopeia to Cygnus. The region is full of easily recognizable asterisms, from the Northern Cross, Cassiopeia's W, and the arrow of Sagitta, to the Square of Pegasus.

Cygnus (the Swan)

Cyg/Cygni; highest at midnight: early August

Flying south along the Milky Way, the swan is a wonderful constellation set against the bright, rich, starry background of the Milky Way. The stargazer can spend literally hours scanning this area through binoculars, sweeping along magnificent star fields and spending a while searching for some of the more elusive deep sky quarry on offer. As well as being one of the three bright stars that make up the Summer Triangle asterism, Alpha Cyg (Deneb) is the top star of the Northern Cross asterism. At the foot of the cross, near the southwestern border of Cygnus, is Beta Cyg (Albireo), one of the most beautiful colored double stars. A small telescope will easily resolve the companion to its golden mag 3.1 primary, a steely blue star of mag 5.1. Omicron Cyg is another wonderful colored double, separable through binoculars, made up of an orange mag 3.8 primary and a sea-green mag 4.8 companion. Closer scrutiny will reveal another companion to the primary, a blue mag 7 star.

M29 is a loose open cluster of around 20 stars, of which several are reasonably bright, is easy to find by sweeping south of Gamma Cyg. M39, a much bigger big open cluster, is made

▲ The North America Nebula in Cygnus, viewed through big binoculars.

up of around a dozen fairly bright stars and many more fainter ones, and makes a lovely low-magnification telescopic sight.

The Blinking Planetary (NGC 6826) lies east of Theta Cyg. It appears small and well-defined, with a slightly blue tinge, its central star visible through a 150mm telescope. Its outer shell only appears to blink off as the observer looks directly at the object, while its bright central star remains visible.

Spread across a portion of the Milky Way in southern Cygnus, the Veil Nebula is a supernova remnant, the brightest parts of which, NGC 6992, may be discerned through big binoculars from a dark site. Big binoculars will also reveal the North America Nebula (NGC 7000), appearing as a wedge-shaped brightening of the Milky Way (considerably larger than the apparent diameter of the full Moon), to the east of Deneb. Having such a large area and a low surface brightness, it is elusive at higher magnifications through a telescope.

N

▲ M29, an open cluster in Cygnus observed through a small telescope.

Lacerta (the Lizard)
Lac/Lacertae; highest at midnight: late August
A narrow, inconspicuous constellation comprising a zigzag of faint stars, Lacerta has one deep sky treasure—NGC 7243, a fairly large open cluster of around 50 sub-8th magnitude stars that resembles a miniature double cluster.

Pegasus (the Winged Horse)
Peg/Pegasi; highest at midnight: early September
Pegasus is a wide constellation, easy to locate because of the prominent Square of Pegasus asterism takes up much of its eastern parts—but note that the top-left star actually belongs to Andromeda. Seeing how many stars are visible within the square is a good test of the darkness of your observing site; if you can spot six stars, then you have a nice dark site.

Epsilon Peg is a lovely wide double star comprising an orange mag 2.4 primary and a blue mag 8.4 companion. Around 4° to its northwest lies M15, a nice globular cluster bright enough to be seen through binoculars; its outer stars are resolvable through a 150mm telescope.

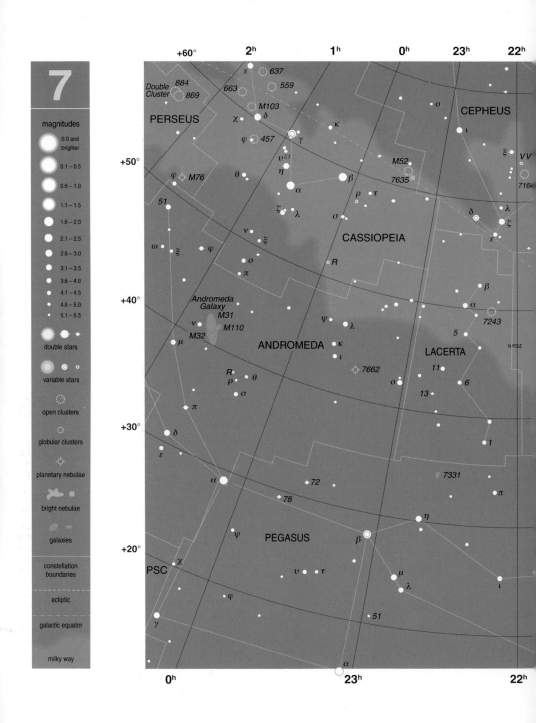

7

magnitudes

- 0.0 and brighter
- 0.1 – 0.5
- 0.6 – 1.0
- 1.1 – 1.5
- 1.6 – 2.0
- 2.1 – 2.5
- 2.6 – 3.0
- 3.1 – 3.5
- 3.6 – 4.0
- 4.1 – 4.5
- 4.6 – 5.0
- 5.1 – 5.5

double stars

variable stars

open clusters

globular clusters

planetary nebulae

bright nebulae

galaxies

constellation boundaries

ecliptic

galactic equator

milky way

+60°   2ʰ   1ʰ   0ʰ   23ʰ   22ʰ

ε   637
Double   884
Cluster   869   663   559
M103
PERSEUS   χ   δ
φ   457   γ   CEPHEUS
ο
υ²,¹   ι
+50°   η   M52
φ   M76   θ   β   7635   ξ   VV
α   716
51   ζ   λ   ρ   τ   δ   λ
ν   ξ   σ   ζ
ω   ξ   φ   CASSIOPEIA   ε
ο   R
π   β
+40°   α
Andromeda   7243
Galaxy   ψ   λ   N. POLE
ν   M31   5
μ   M32   M110   κ   LACERTA
ANDROMEDA   ι   11
R   θ   7662   6
ρ   ο   13
σ   1
π
+30°   δ
ε   7331
π
α   72
78   η
PSC   χ   ψ   PEGASUS   β   μ
+20°   υ   τ   λ   ι
φ
γ   51
α
0ʰ   23ʰ   22ʰ

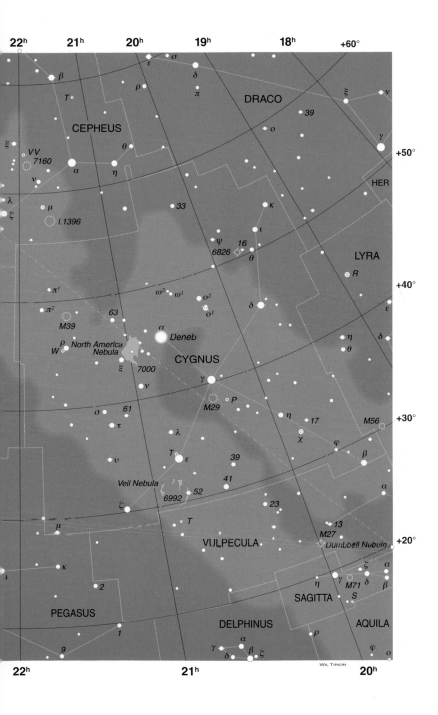

WIL TIRION

▶ # Chart 8

Containing few bright stars, with no trace of the Milky Way
and having a meager selection of bright deep sky objects,
this area along the celestial equator is one of the sparsest
regions of the heavens.

▲ Binoculars will show little
Aries at its best.

Aries (the Ram)
Ari/Arietis; highest at midnight: early November

Aries, the smallest constellation of the zodiac,
can be identified by the small pattern of its
brighter stars—Alpha (Hamal), Beta, and
Gamma Ari—which lie some distance west of
the Pleiades in neighboring Taurus.

Gamma Ari is a double of white mag 4.6
stars, easily visible through small telescopes,
looking like a pair of cat's eyes. Lambda Ari is
another wide double, with a white mag 4.8
primary and a yellow mag 7.3 companion. The
constellation harbors no conspicuous deep
sky objects.

Cetus (the Whale)
Cet/Ceti; highest at midnight: mid-October

A very large constellation lying south of the
ecliptic, the bulk of the whale is immersed
south of the celestial equator, but its head rears
north and eyes little Aries and glamorous
Taurus. Its second brightest star, the red giant
Alpha Cet (Menkar), makes a wide but
charming double with the blue 5th mag star 93
Cet to its north. Beta Cet (mag 2), far to the
southwest, is the brightest star in Cetus. Two
stars of particular interest to astronomers can
be seen with the unaided eye. The red giant
Omicron Cet (Mira), is a favourite of variable
star observers. It varies between the 3rd and
9th mags over a period of around 332 days.

Tau Cet is a star remarkably like our own Sun; just 12 light years away, it has an apparent brightness of mag 3.5. Observations have revealed that the space surrounding Tau Cet contains ten times more asteroid and comet debris than found in the Sun's vicinity; any planets there are likely to experience high rates of catastrophic collisions.

The 8th magnitude planetary nebula NGC 246 makes a nice target for close scrutiny with a 200mm telescope. At high magnification it appears as a gray mottled ellipse, with several 11th magnitude stars visible in its vicinity, including its central star. M77, a 10th magnitude spiral galaxy with a bright, highly condensed nucleus, can be found by sweeping less than one degree east of Delta Cet.

Pisces (the Fishes)

Psc/Piscium; highest at midnight: early October

Lying to the immediate south and east of the Square of Pegasus, Pisces is one of the largest of the twelve zodiacal constellations. Its traditional outline comprises a series of faint stars, and it is only traceable from dark sites. Pisces' brightest star, Alpha Psc, lies in the far southeastern corner of the constellation. Through a 100mm telescope it can be resolved as a close double of mags 4.2 and 5.2. In the western corner of Pisces, the well-known asterism of the Circlet is made up of seven stars —a challenge to spot with the unaided eye from an urban site.

The Phantom (M74), a face-on spiral galaxy, is Pisces' brightest deep sky object. It can be found a little more than one degree east of Eta Psc, and appears as a sizeable round smudge with a bright, well-defined nucleus through a small telescope.

▼ The face-on spiral galaxy the Phantom (M74) is easy to locate through a small telescope.

N

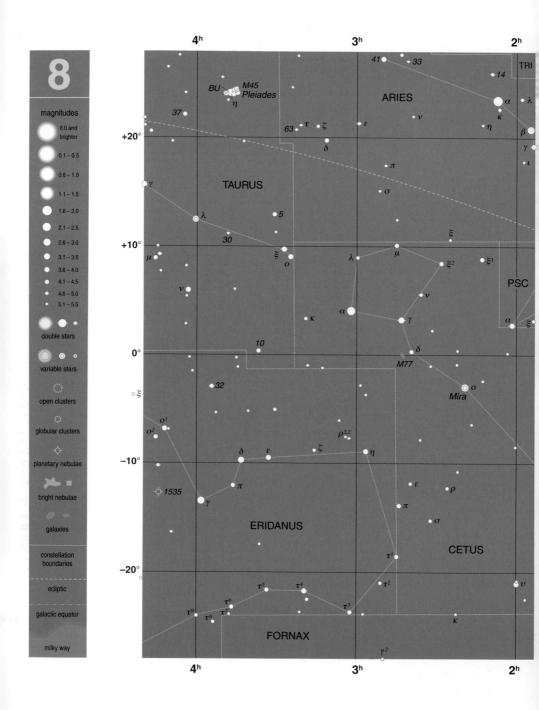

8

magnitudes

- 0.0 and brighter
- 0.1 – 0.5
- 0.6 – 1.0
- 1.1 – 1.5
- 1.6 – 2.0
- 2.1 – 2.5
- 2.6 – 3.0
- 3.1 – 3.5
- 3.6 – 4.0
- 4.1 – 4.5
- 4.6 – 5.0
- 5.1 – 5.5

double stars

variable stars

open clusters

globular clusters

planetary nebulae

bright nebulae

galaxies

constellation boundaries

ecliptic

galactic equator

milky way

4ʰ   3ʰ   2ʰ

TRI

41   33   14

ARIES

BU   M45
Pleiades

37   η

63   τ   ζ   ε   ν   η   κ   α   λ
+20°   β
δ   γ
π   ι

γ   σ

TAURUS

λ   5

30   ξ
+10°   ξ¹
μ   ξ   o   λ   μ   ξ²
ν   ν
κ   α   γ   α   ξ

PSC

10   δ
0°   M77
32   o
ξ   Mira
o¹
o²   ρ³,²
δ   ε   ζ   η   ε   ρ
1535   π   π
−10° γ   σ
ERIDANUS
CETUS
τ¹
−20°   τ²
τ⁵   τ⁴   υ
τ⁶   τ³
τ⁹ τ⁸ τ⁷   κ
FORNAX
γ²

4ʰ   3ʰ   2ʰ

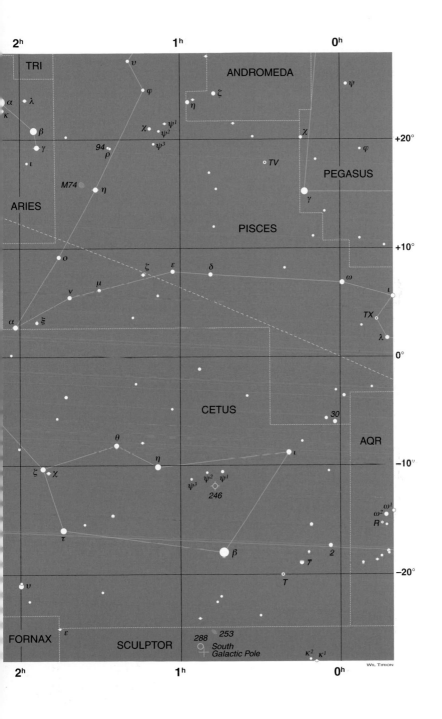

2ʰ   1ʰ   0ʰ

TRI   ANDROMEDA
                                                        ψ
                              υ
                         φ   ζ
                                   η
α   λ
κ   χ   ψ¹   χ
    β   ψ²   +20°
                                   ψ³   φ
   γ   94   PEGASUS
  ι   ρ   TV
                                                     γ
       M74   η   PISCES
ARIES

                                                                  +10°
         o   ω
                    ζ   ε   δ   ι
      γ   μ   TX
α   λ
    ξ   0°

                              CETUS
                                                    30
                                                    AQR
         θ
              η   −10°
    ζ   χ   ψ³ ψ² ψ¹
                         246
                                                        ω¹
                                                    ω²
                                                    R
    τ   2
                        β   7   −20°
   υ   T

FORNAX   ε
         SCULPTOR   288   253
                         South   κ² κ¹
                         Galactic Pole
                                        WIL TIRION
2ʰ   1ʰ   0ʰ

▶ Chart 9

Plenty of dazzling stars and bright deep sky objects are spread throughout this region, centered around Orion. Probably the most exciting part of the heavens, there's more than enough to hold the attention of stargazers of any level for hours on end.

Canis Minor (the Little Dog)

CMi/Canis Minoris; highest at midnight: early January

A small constellation, but one that most stargazers are familiar with, since its brightest star Alpha CMi (Procyon) is one of the brightest in the skies.

Lepus (the Hare)

Lep/Leporis; highest at midnight: mid-December

Located at the feet of Orion, Lepus is a fairly bright constellation, and its broad bowtie pattern can easily be traced from the USA. A small telescope will reveal Gamma Lep to be a double star, comprising a yellow mag 3.6 primary and an orange mag 6.2 companion.

NGC 2017, a lovely group of stars, lies 1.5° east of Alpha Lep. A small telescope will show six stars, four of which stand out. In the south of Lepus, set in a rich star field, the compact globular cluster M79 is a superb sight.

Monoceros (the Unicorn)

Mon/Monocerotis; highest at midnight: early January

Monoceros is a faint, sprawling constellation. Beta Mon is a dazzling triple, comprising blue stars of mag 3.8, 5 and 5.3, and easily resolvable through a small telescope; 8 Mon is a line-of-sight double comprising a yellow mag 4.4 and blue mag 6.7 star.

S Mon, a bright blue star with a mag 7.6 companion, lies within the Christmas Tree Cluster (NGC 2264), easily visible through binoculars. To the southwest lies the Rosette Nebula, the faint backdrop for the bright star cluster NGC 2244. It is possible to glimpse the Rosette through big binoculars, but it may not

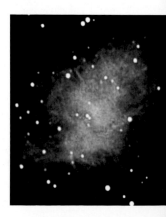

▼ The Crab Nebula, a supernova remnant in Taurus.

be obvious through a larger telescope at higher magnifications.

M50 is a sizeable grouping of around 30 fairly bright stars, with an equal number of fainter ones, easily resolvable through a 150mm telescope.

Orion (the Hunter)
Ori/Orionis; highest at midnight: mid-December

Orion is the most magnificent constellation. Straddling the celestial equator, its shape is recognizable to stargazers in both hemispheres.

Sigma Ori is a lovely multiple; its main star of mag 3.8 has two nearby 6th magnitude stars, and a more distant triple of 8th magnitude stars.

A short distance south of Orion's Belt, a misty patch can be discerned with the keen unaided eye. This is the Orion Nebula (M42), one of the biggest and brightest nebulae in the heavens. Considerable structure can be seen with binoculars, and a small telescope will reveal a glowing greenish mass with delicate wisps, intruded upon by a prominent dark lane. Several stars can be seen in and around the nebula, notably the Trapezium (Theta Ori), a bright quadruple star.

▲ Orion climbs high above the Alps.

Taurus (the Bull)
Tau/Tauri; highest at midnight: early December

Brooded over by the red bullseye of Alpha Tau (Aldebaran), this large and easily identifiable zodiacal constellation dominates the skies northwest of Orion. Aldebaran is set against the background of the Hyades, a large V-shaped open cluster containing a dozen or more naked eye stars in an area that can be covered by a clenched fist.

In the northwest of Taurus, a handful of the brightest stars within the Pleiades star cluster (M45) is easy to see with the unaided eye. Telescopes will reveal dozens of young blue stars, and near the Pleiad 23 Tau (Merope) may be seen a hint of reflecting nebulosity.

Just over 1° north of Zeta Tau, the faintly glowing supernova remnant of the Crab Nebula (M1) requires an 80mm telescope to be seen. Through large instruments, it appears as a gray elliptical patch.

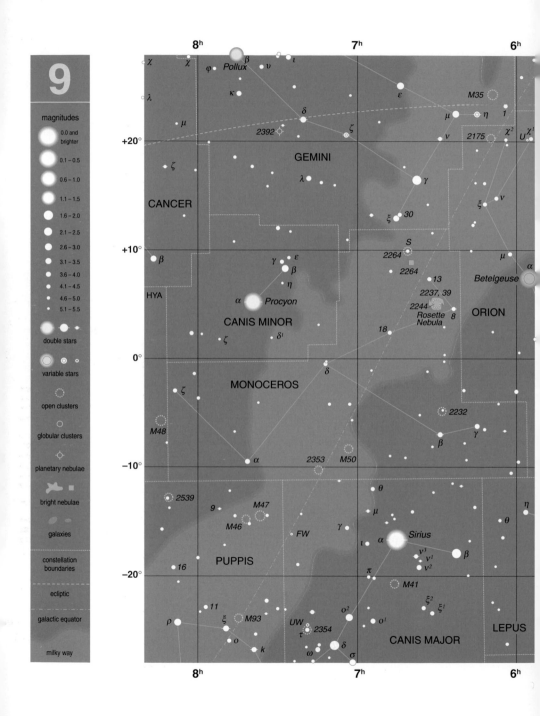

9

magnitudes

- 0.0 and brighter
- 0.1 – 0.5
- 0.6 – 1.0
- 1.1 – 1.5
- 1.6 – 2.0
- 2.1 – 2.5
- 2.6 – 3.0
- 3.1 – 3.5
- 3.6 – 4.0
- 4.1 – 4.5
- 4.6 – 5.0
- 5.1 – 5.5

double stars

variable stars

open clusters

globular clusters

planetary nebulae

bright nebulae

galaxies

constellation boundaries

ecliptic

galactic equator

milky way

χ χ φ *Pollux* β υ ι
λ κ ε *M35*
μ δ μ η 1
2392 ζ ν 2175 χ² U χ¹
+20° ζ **GEMINI** ξ ν
CANCER λ γ
ξ 30
β γ ε S μ
+10° β β 2264 α
η 2264 13 *Betelgeuse*
HYA α *Procyon* 2237, 39
2244 8 **ORION**
CANIS MINOR δ¹ Rosette Nebula
ζ 18
0° δ
ζ **MONOCEROS**
2232
M48 2232
β γ
α 2353 *M50* β γ
−10° α
θ
2539 η
9 *M47* μ θ
M46 γ
FW *Sirius*
16 ι α ν³ ν¹ β
π ν²
−20° *M41*
11 ξ² ξ¹
ρ ξ *M93* o²
o UW 2354 o¹ **LEPUS**
τ
k ω δ **CANIS MAJOR**
σ
8ʰ **7ʰ** **6ʰ**

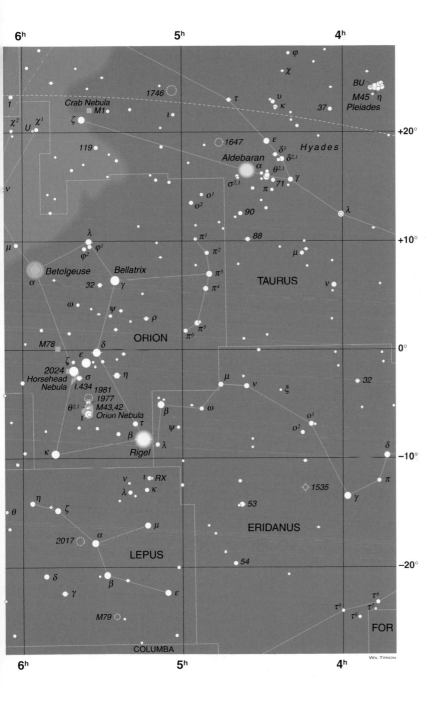

6ʰ 5ʰ 4ʰ

1

χ²
U χ¹
ν

μ

Crab Nebula
M1

1746

119

λ
φ¹
φ²

Betelgeuse
α

Bellatrix

32
γ

ω
ψ

ρ

ORION

M78

2024
Horsehead
Nebula

ζ ε
δ
σ
η
I.434 1981
1977
θ²·¹ M43,42
ι Orion Nebula
β
ψ
τ
λ
β
κ Rigel

ν ι RX
λ κ

η ζ

2017 α μ

LEPUS

δ
γ β ε

M79

COLUMBA

φ
χ

BU
M45 η
Pleiades

τ
υ
κ
37

1647
ε
δ³
Aldebaran δ²·¹
α θ²·¹ Hyades
σ²·¹ γ
π 71
o¹
o²
90
88
π¹
π²
μ
π³
TAURUS
ν
π⁴

π⁵
π⁶

λ

μ
ξ ν
ω o¹
o²
32

δ

π

γ

1535
53
ERIDANUS
54
1746

τ⁹
τ⁶
τ⁷
τ⁸

FOR

+20°

+10°

0°

−10°

−20°

6ʰ 5ʰ 4ʰ

WIL TIRION

▶ Chart 10

An interesting region bordering the celestial equator, containing few bright stars but populated with a number of splendid star clusters and galaxies. The ecliptic slices through the area, from Virgo in the east, through Leo and Cancer, to Gemini in the far northwestern corner of the chart.

Cancer (the Crab)
Cnc/Cancri; highest at midnight: early February

Cancer is the least prominent of all the 12 zodiacal constellations, but it can easily be located southeast of the stellar twins Castor and Pollux.

Small telescopes will split the two main components of Zeta Cnc, widely separated stars of mag 5.2 and 5.8. A 200mm telescope will reveal that the brighter star has a closer mag 6.2 companion. Iota Cnc is a lovely colored double, with a yellow mag 4 primary and a blue mag 6.6 companion, easy to resolve with a small telescope.

Keen eyes will discern a misty patch just north of Delta Cnc. Binoculars reveal this to be the Beehive Cluster (M44), a sizeable star swarm covering an area of around ten times that of the full Moon. A very low telescopic magnification will take in the entire cluster.

Less than 2° west of Alpha CnC lies a much smaller open cluster, the King Cobra (M67). M67 is a major open cluster in its own right, containing 300 stars and appearing as a full Moon-sized oval smudge through binoculars.

Crater (the Cup)
Crt/Crateris; highest at midnight: mid-March

Located south of the celestial equator, Crater is a fairly inconspicuous group of stars.

Hydra (the Water Snake)
Hya/Hydrae; highest at midnight: March

Covering more than 1,300 square degrees, Hydra is the largest constellation. Hydra's most southerly stars only just manage to clear the horizon from southern UK latitudes, making it difficult to trace in its entirety.

Epsilon Hya is a nice colored double, a binary with a yellow mag 3.4 primary and a blue mag 6.7 partner, resolvable with a small telescope. On the opposite side of the constellation, deep in the south, another notable colored double 54 Hya, has a yellow mag 5.1 primary and a lilac mag 7.2 companion.

The Lawn Sprinkler (M48) is a large open cluster, easily visible through binoculars. Telescopes will show a spray of fairly bright stars, some of which appear as close pairs. M68 is a visually interesting bright globular cluster, with intriguing dark lanes and a sizeable dark notch visible through a 200mm telescope. To its east lies the Southern Pinwheel (M83), a face-on barred spiral galaxy, easy to see through binoculars. Its spiral shape can be seen through a 150mm telescope.

Small telescopes will show the glowing turquoise disk of the Ghost of Jupiter (NGC 3242), one of the nicest planetary nebulae. A 150mm telescope will show its 11th magnitude central star.

Leo (the Lion)
Leo/Leonis; highest at midnight: early March
Leo dominates its immediate neighborhood. It is easy to find, with the prominent Sickle asterism in the west and a bright triangle of stars forming its tail in the east.

Alpha Leo (Regulus) at the base of the Sickle is a bright double star whose components of mag 1.4 and 7.7 can be easily separated through a small telescope. Gamma Leo is another charming double, a pair of yellow stars of mag 2.3 and 3.6.

M65 and M66, 9th magnitude spiral galaxies, are less than half a degree apart and can be viewed through a low-power field. Another notable triplet of galaxies lies a short distance to the west, and comprises M95, M96, and M105.

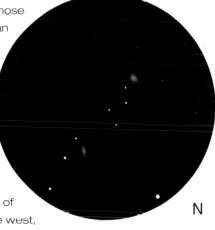

N

▲ M65 and M66, a close bright galactic duo in Leo.

Sextans (the Sextant)
Sex/Sextantis; highest at midnight: late February
Boxed-in south of Leo, Sextans is one of the least conspicuous constellations. It has one notable feature, the Spindle Galaxy (NGC 3115), a 9th magnitude galaxy that resembles a glowing eye with a bright pupil through a 150mm telescope.

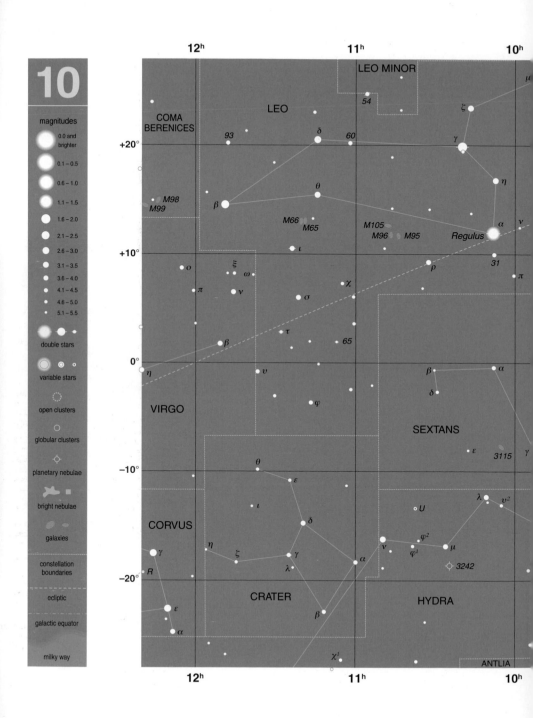

10

magnitudes

- 0.0 and brighter
- 0.1 – 0.5
- 0.6 – 1.0
- 1.1 – 1.5
- 1.6 – 2.0
- 2.1 – 2.5
- 2.6 – 3.0
- 3.1 – 3.5
- 3.6 – 4.0
- 4.1 – 4.5
- 4.6 – 5.0
- 5.1 – 5.5

double stars

variable stars

open clusters

globular clusters

planetary nebulae

bright nebulae

galaxies

constellation boundaries

ecliptic

galactic equator

milky way

12ʰ   11ʰ   10ʰ

LEO MINOR

LEO

COMA BERENICES

54

93   δ   60   γ

+20°   ζ   μ

θ   η

β   M98   M99

M66   M105

M65   M96   M95   α

ι   Regulus   ν

+10°   ρ   31   π

o   ξ   ω

π   ν   χ

σ

β   τ

65

0°   η   υ   φ

β   α

δ

VIRGO   SEXTANS

ε   3115

–10°   θ   γ

ε

ι   U   λ   υ²

δ   φ²

CORVUS   ν   φ¹   μ

γ   η   ζ   γ   3242

R   λ   α

–20°   CRATER   HYDRA

ε   β

α   χ¹   ANTLIA

12ʰ   11ʰ   10ʰ

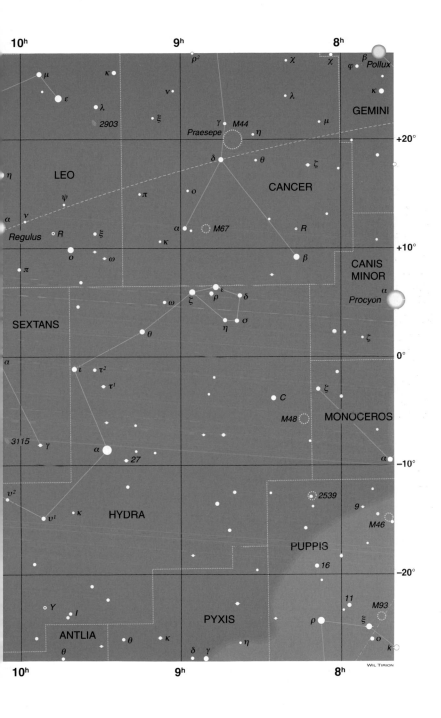

ρ²   χ   χ   φ   β   Pollux

μ   κ   ν   λ   κ   GEMINI

ε   λ   ξ   μ

2903   γ   M44

Praesepe   η

+20°

δ   θ   ζ

η   LEO   CANCER

ψ   π   o

α   γ   α   M67   R

Regulus   R   ξ   +10°

κ

π   o   ω   β   CANIS
MINOR

SEXTANS   ω   ζ   ρ   δ   α
Procyon

θ   η   σ

α   ζ   0°

ι   τ²

τ¹

C   ζ

3115   M48   MONOCEROS

γ   α   27   α   −10°

υ²   2539

9

υ¹   κ   HYDRA   M46

PUPPIS

16   −20°

11   M93

Υ   ι   ξ

ANTLIA   θ   κ   ρ   o

θ   PYXIS   η   k

δ   γ

Wil Tirion

▶ Chart 11

Two stars in this region are particularly prominent—orange Arcturus in Boötes and ice-white Spica in Virgo. Virgo takes up much of this region along the celestial equator. A mass of bright galaxies spreads south from Virgo into Coma Berenices (see page 99, A galaxy hunt).

Coma Berenices (Berenice's Hair)
Com/Comae Berenices; highest at midnight: early April

▲ The Black Eye Galaxy, M64, in Coma Berenices.

Coma is an inconspicuous constellation made up of around 30 faint stars on the border of naked eye visibility. The eye's attention is drawn to an extensive misty patch, the Coma Star Cluster (Melotte 111) containing around 40 faint stars spread over a 5° wide area.

Hundreds of stars within M53, a bright globular cluster with a concentrated core, can be easily viewed through a 200mm telescope at high magnification. M53 is easy to locate, just one degree northeast of Alpha Com.

All the brighter galaxies of Coma lie within the Coma-Virgo Cluster (described on page 99), a vast assembly of galaxies around 50 million light years away. Probably the most spectacular is the Black Eye Galaxy (M64), a bright galaxy whose prominent silhouetted dark lane is easily visible through a 150mm telescope. There are many more Coma galaxies easily visible through small telescopes, including M85, M88, M99, and M100.

Corvus (the Crow)
Crv/Corvi; highest at midnight: early April

Corvus is a small constellation, recognizable in a dark sky by its four brightest stars which form a small trapezium to the west of Spica. Delta Crv forms an easy naked eye double with Eta Crv. Small telescopes reveal that Delta Crv is a double star with a white mag 2.9 primary and a fainter violet mag 9.4 companion.

Libra (the Scales)

Lib/Librae; highest at midnight: mid-May

One of the smallest and least conspicuous of the zodiacal constellations, Libra's main stars form a quadrilateral that straddles the ecliptic. It is located northwest of Scorpius, but it is a difficult constellation to see with the unaided eye from a city.

Alpha Lib (Zubenelgenubi) is a wide double separable in binoculars, consists of a mag 2.7 sky blue star with a mag 5.2 white companion. Beta Lib is the brightest of Libra's stars, and displays an uncommon green hue, the color being particularly notable through binoculars.

Serpens Caput (the Serpent's Head)

Ser/Serpentis; highest at midnight: mid-May

The brighter half of Serpens is made up of a line of fairly bright stars sandwiched between Hercules and Virgo, from Corona in the north to a few degrees south of the celestial equator.

M5 is one of the best globular clusters in the entire sky. Just visible from a dark site with the naked eye, it is easily spotted through binoculars and a joy to view close-up through a telescope. Measuring around two-thirds the apparent diameter of the full Moon, M5 contains lots of bright stars resolvable through a 150mm telescope.

Virgo (the Virgin)

Vir/Virginis; highest at midnight: mid-April

Virgo, the sky's second biggest constellation, occupies almost 1,300 square degrees of sky. Gamma Vir (Porrima) is a famous double star of equal components, both white stars shining at mag 4.6. The pair orbit one another in a period of around 170 years, and are closest in 2005, when they will be separable at high magnification using a 250mm telescope. All of the bright galaxies in Virgo belong to the Coma-Virgo Cluster (see page 99). Most of them lie in the northwest of the constellation, and include M58, M59, M60, the Swelling Spiral (M61), M84, M87, and M90. In the southwest of Virgo can be found the Sombrero Hat (M104), a bright 8th magnitude edge-on galaxy, whose nuclear bulge rises smoothly on either side of its spindly spiral arms.

▶ The Sombrero Hat, M104, an edge-on galaxy in Virgo. **N**

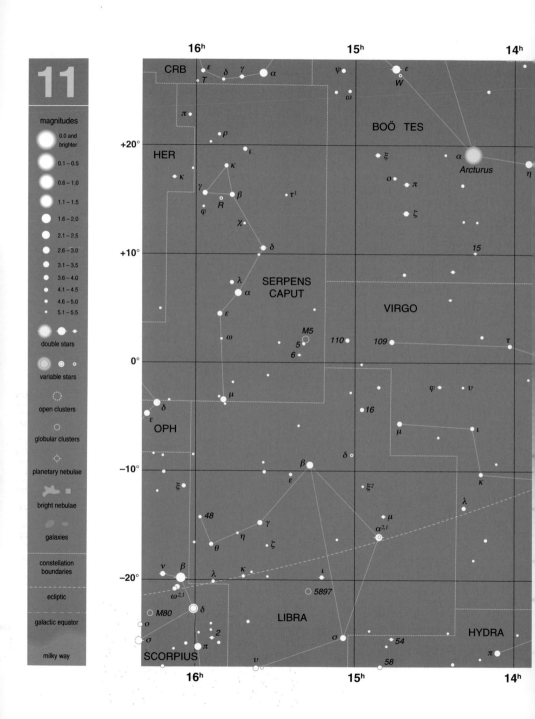

11

magnitudes

- 0.0 and brighter
- 0.1 – 0.5
- 0.6 – 1.0
- 1.1 – 1.5
- 1.6 – 2.0
- 2.1 – 2.5
- 2.6 – 3.0
- 3.1 – 3.5
- 3.6 – 4.0
- 4.1 – 4.5
- 4.6 – 5.0
- 5.1 – 5.5

double stars

variable stars

open clusters

globular clusters

planetary nebulae

bright nebulae

galaxies

constellation boundaries

ecliptic

galactic equator

milky way

16ʰ   15ʰ   14ʰ

CRB   ε δ γ α   ψ   ε
       T   ω   W

       π   BOÖTES

ρ

+20°   ι

HER   ξ   α
       κ   Arcturus
κ   ο   π   η
   γ κ
   φ β   τ¹
      R   ζ

χ

+10°   δ   15

       λ   SERPENS
       α   CAPUT   VIRGO

       ε
       ω   M5
                 5   110   109   τ
0°   6

                              φ   υ

μ   16

OPH δ   μ   ι
     ε
                    δ

−10°   β
            ε   ξ²   κ
       ξ   λ
   48   γ   μ
       η   ζ   α²·¹
       θ

   ν β   ι
−20°   λ   κ
   ω²·¹   5897
   M80 δ   LIBRA   HYDRA
   ο   σ
σ   π   54   π
SCORPIUS   υ   58
16ʰ   15ʰ   14ʰ

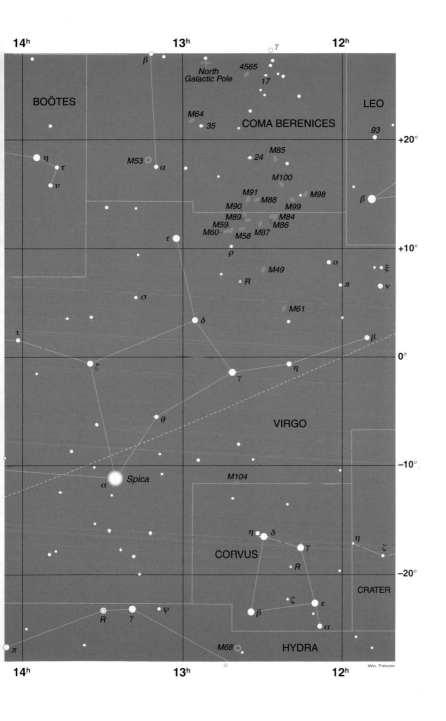

γ

BOÖTES

β

North
Galactic Pole

4565

17

LEO

COMA BERENICES

M64

35

93

+20°

M53

α

24

M85

η

τ

υ

M100

M91

M88

M98

M90

M99

M89

M84

M59

M86

M60

M58

M87

ε

ρ

+10°

M49

o

R

ξ

π

ν

σ

M61

δ

β

0°

ζ

γ

η

τ

θ

VIRGO

−10°

α Spica

M104

CORVUS

η

δ

η

γ

ζ

R

−20°

ξ

ε

CRATER

β

R

γ

ψ

α

π

M68

HYDRA

Wil Tirion

STAR CHARTS & SHOWPIECES

149

▶ # Chart 12

As the Milky Way plunges south from Aquila into Sagittarius, a prominent rift can be traced along its middle, caused by vast clouds of gas and dust silhouetted against the distant starry background.

Aquila (the Eagle)

Aql/Aquilae; highest at midnight: mid-July

▲ M16 star cluster and the Eagle Nebula in Serpens Cauda.

Aquila, a medium-sized constellation, straddles the celestial equator. Aquila contains two nicely colored double stars, easily visible through small telescopes: 15 Aql, a mag 5.4 orange star with a lilac mag 7 companion; and 57 Aql, a sky blue mag 5.7 primary with a mag 6.5 companion. Aquila contains a fair scattering of faint planetary nebulae, but its deep sky showpiece is NGC 6709, an open cluster made up of around 30 fairly bright stars, a number of which are arranged in loose chains.

Ophiuchus (the Serpent Bearer)

Oph/Ophiuchi; highest at midnight: mid-June

A large constellation that extends well south of the celestial equator, Ophiuchus' main stars are of the 2nd and 3rd magnitude. Its full outline can be hard to trace from northern climes, since it extends to 30° S.

The multiple star Rho Oph makes a great high magnification sight. It consists of a mag 4.6 primary and close mag 5.7 partner, plus two more widely separated outlying stars of 7th magnitude.

North of Beta Oph lies IC 4665, a scattered open cluster visible with the naked eye as a misty patch, best viewed through binoculars.

Ophiuchus is rich in globular clusters, seven of the brightest appearing in Messier's list: M9, M10, M12, M14, M19, M62, and M107. The brightest of these, M10 and M12, can be resolved through a 200mm telescope.

Sagitta (the Arrow)

Sge/Sagittae; highest at midnight: mid-July

Immediately south of Cygnus, buried in the Milky Way, lies the diminutive

Sagitta, whose main stars are arranged like a feathered arrow. Just below the arrow's mid-shaft lies M71, a bright globular cluster which appears almost homogenous in density, set in a rich star field.

Scutum (the Shield)
Sct/Scuti; highest at midnight: Early July

Scutum may be tiny, and its stars difficult to discern with the unaided eye, but it is packed with deep sky wonder. Set against the gorgeous Milky Way background, the Wild Duck (M11) is a glorious broad arc of around 200 stars; visible through binoculars as a hazy patch, the cluster is resolvable through a 100mm telescope at higher magnifications. A few degrees to its south lies a fainter and less populous open cluster, M26.

Serpens Cauda (the Serpent's Tail)
Ser/Serpentis; highest at midnight: late June

Serpens is a constellation split into two sections on either side of Ophiuchus. Serpens Cauda, the constellation's fainter half, is made up of a line of stars that run along the Milky Way's dark central reservation. M16, an open star cluster comprising a scattering of around a dozen stars, can be discerned through binoculars.

Vulpecula (the Fox)
Vul/Vulpeculae; highest at midnight: late July

Vulpecula straddles the Milky Way. It contains no bright stars, and it is likely to be invisible to the unaided eye from an urban location. Brocchi's Cluster (Collinder 399) is a neat binocular asterism made up of 5th magnitude stars. It is often called the Coathanger, though binoculars show it upside down from the northern hemisphere. The showpiece of Vulpecula, however, is the Dumbbell Nebula (M27), the finest planetary nebula in the northern skies. Easily visible as a large, well-defined patch through binoculars, M27 resembles a broad misty bow tie; its greenish hue is noticeable through larger telescopes.

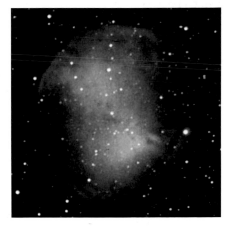

▶ M27, the Dumbbell Nebula in Vulpecula.

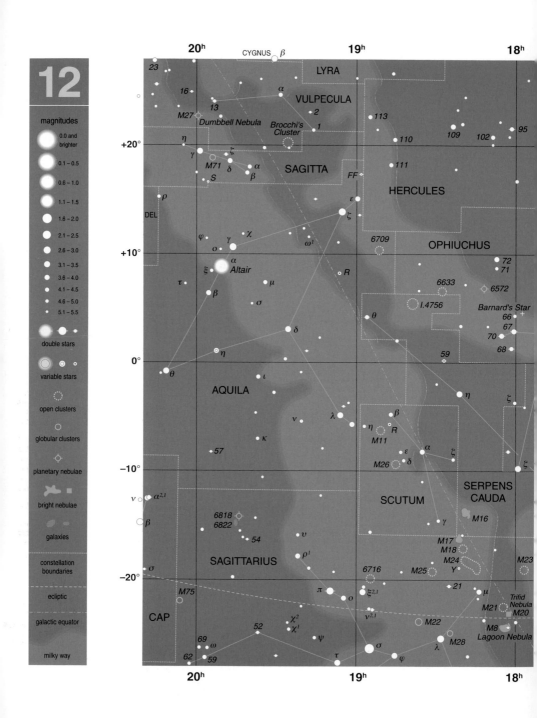

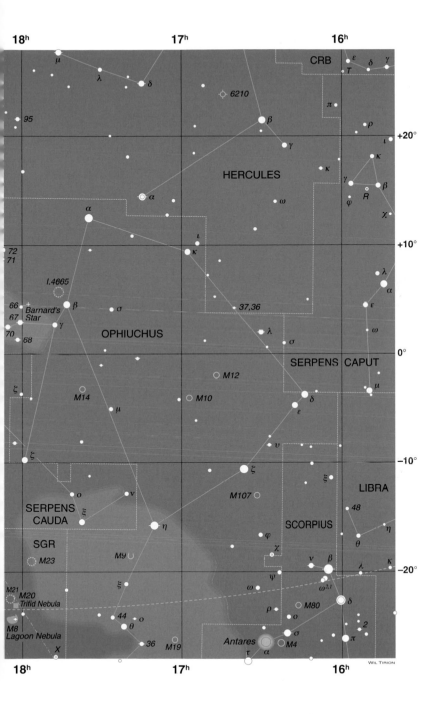

WIL TIRION

▶ Chart 13

This area along the celestial equator is crossed by the ecliptic, which runs from western Pisces, through Aquarius and Capricornus. Having few bright stars and being short of prominent deep sky objects, it is a generally unspectacular region. The delightful little constellations of Sagitta (see Chart 12) and Delphinus lie in the northwest.

Aquarius (the Water Carrier)

Aqr/Aquarii; highest at midnight: late August

Aquarius is a large constellation whose main pattern comprises a widely-spaced collection of 2nd and 3rd magnitude stars lying south of the celestial equator. Immediately east of Alpha Aqr, a small but striking asterism often called the Propeller or the Water Jar comprises a pattern of four stars, whose central star, Zeta Aqr, is a binary with twin white components of mag 4.3 and 4.4.

Three different types of deep sky object—a planetary nebula, an open cluster, and a globular cluster, all visible in the same field of view at low magnifications—can be found in the western reaches of Aquarius. Less than 1.5° west of Nu Aqr lies the Saturn Nebula (NGC 7009), a superb planetary nebula, visible through an 80mm telescope as an elliptical disk of a similar size to Saturn. To its southwest is M73, a small open cluster shaped rather like the Propeller asterism. Nearby M72, a small 9th magnitude globular cluster with a bright core, can be difficult to resolve using anything smaller than a 250mm telescope.

M2, the brightest globular cluster in the region, is located less than 5° north of Beta Aqr. It can be seen through binoculars as a fuzzy patch, and a large number of its stars can be seen through a 150mm telescope.

In the far south of Aquarius, the Helix Nebula (NGC 7293) can be seen fairly easily through binoculars as a circular smudge. Although it has the largest apparent diameter of all planetary nebulae, it has a low surface brightness, so the best views are at low magnifications.

Capricornus (the Sea Goat)

Cap/Capricorni; highest at midnight: early August

A fairly small zodiacal constellation, and quite an obscure one too, Capricornus comprises a collection of 3rd and 4th magnitude stars arranged in a broad south-pointing arrowhead pattern. Alpha Cap is a close naked eye double comprising Alpha 2 Cap, a yellow giant of mag 3.6 and Alpha 1 Cap, an orange supergiant of mag 4.3. The pair are a line-of-sight double. Each of these stars is itself a wide double, with faint companions, the dimmest visible through a 200mm telescope. Beta Cap is a nice colored double with a golden mag 3 primary and a sky-blue mag 6.1 partner, wide enough to be seen through binoculars.

Capricornus' best deep sky delight is M30, a middling sized, fairly bright globular cluster whose outer regions can be resolved through a 200mm telescope.

▲ The small but beautiful constellation of Delphinus.

Delphinus (the Dolphin)

Del/Delphini; highest at midnight: early August

Gamma Del, the dolphin's beak, is a colored double consisting of a golden mag 4.3 and a lemon-yellow 5.2 star, which are easily separable using just small telescopes. The tail points south to NGC 6934, a much-neglected globular; small and bright, it has a concentrated core but it is very difficult to resolve through anything smaller than a 300mm telescope.

Equuleus (the Little Horse)

Equ/Equulei; highest at midnight: mid-August

Covering just 72 square degrees, this inconspicuous constellation ranks as the second smallest in the heavens. Equuleus has no stars brighter than mag 3.9, and it is difficult to identify with the naked eye from light-polluted urban sites. Binoculars will reveal the five main stars set amid dozens of much fainter ones, by simply sweeping one dolphin's length southeast of Delphinus. The main stars can be encompassed within the field of view of 7x50 binoculars. Gamma Equ makes a binocular double with 6 Equulei to its south.

The 1 Equ is an easy telescopic double comprising components of mag 5.2 and 5.9. The fainter component is itself a close double, currently closing and resolvable through a 200mm telescope.

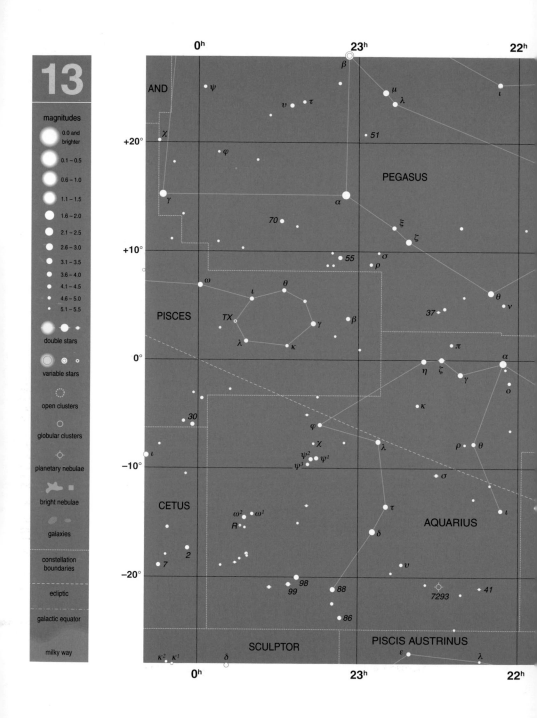

13

magnitudes

- ⬤ 0.0 and brighter
- ⬤ 0.1 – 0.5
- ⬤ 0.6 – 1.0
- ⬤ 1.1 – 1.5
- ⬤ 1.6 – 2.0
- ⬤ 2.1 – 2.5
- ⬤ 2.6 – 3.0
- ⬤ 3.1 – 3.5
- ⬤ 3.6 – 4.0
- • 4.1 – 4.5
- • 4.6 – 5.0
- · 5.1 – 5.5

double stars

variable stars

open clusters

globular clusters

planetary nebulae

bright nebulae

galaxies

constellation
boundaries

ecliptic

galactic equator

milky way

AND

PEGASUS

PISCES

CETUS

AQUARIUS

SCULPTOR

PISCIS AUSTRINUS

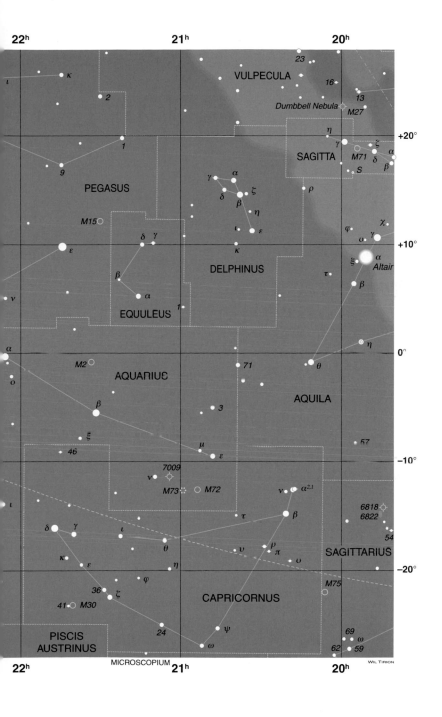

22ʰ   21ʰ   20ʰ

ι   κ

2

VULPECULA

23

16

Dumbbell Nebula   13

M27

PEGASUS

9

1

η   +20°

γ   ζ

SAGITTA   M71 δ   α

S   β

γ α

δ ζ

β η

ι ε

κ

M15

DELPHINUS

δ γ

M15

β

α

EQUULEUS

1

ρ

φ

χ

γ

υ   +10°

ξ   α

Altair

τ

β

ν

η   0ⁿ

α

ο

M2

AQUARIUS

71

θ

β

3

AQUILA

ξ

46

μ

ε

57

7009

ν

M73   M72

ν   α²⁾¹

6818

6822

ι

τ

β

54

δ   γ

ι

θ   υ

ρ π

υ

SAGITTARIUS

κ

ε

η

−20°

φ

M75

36

ζ

CAPRICORNUS

69

41   M30

24

ψ

62   59

PISCIS

AUSTRINUS

ω

ω

▶ Chart 14

Eridanus meanders through an area that is poor in bright
stars and somewhat lacking in bright deep sky objects. We
peer in the direction of the south galactic pole, so the Milky
Way is absent from this chart. The single stellar highlight of
the region depicted is Eridanus' brightest star, Achernar.

Eridanus (the River)
Eri/Eridani; highest at midnight: November

Eridanus is a large constellation whose stars wind from the southern
border of Taurus to the northern border of Hydrus. Much of northern
Eridanus—indeed, the bulk of its actual area—may be viewed from far
northern climes. The whole of the constellation may only be viewed from
locations south of 32° N.

Mag 0.5 Alpha Eri (Achernar) in the far south is easily the brightest of
the constellation's stars. Theta Eri is a lovely blue double of mag 3.2 and
4.3, easily separable through a small telescope. Eridanus' nicest colored
double is 32 Eri, a wide double consisting of a yellow mag 4.8 star and a
turquoise 6.1 companion.

Despite occupying an area of more than 1,100 square degrees, there
is a dearth of bright deep sky objects within Eridanus. Cleopatra's Eye
(NGC 1535), a lovely bright planetary nebula located 4° east of Gamma
Eri, is sufficiently north to be fully appreciated from the USA. It appears as
a sizeable 9th magnitude green-hued ball through a small telescope,
and some structure, including its central star, is discernible through a
150mm telescope.

Fornax (the Furnace)
For/Fornacis; highest at midnight: early November

A sizeable but inconspicuous constellation wedged into the western bend
of river Eridanus. Alpha For is a nice double with a white mag 3.9 primary
and yellow mag 7.2 companion, divisible through a small telescope. The
constellation's best deep sky object is NGC 1365, a 10th magnitude
barred spiral.

Horologium (the Pendulum Clock)
Hor/Horologii; highest at midnight: early November

Practically untraceable to all but the most experienced of stargazers,

Horologium consists of a scattered bunch of stars on the border of naked eye visibility. The constellation's deep sky saving grace is NGC 1261, a bright 8th magnitude globular cluster visible through binoculars.

Reticulum (the Net)
Ret/Reticulum; highest at midnight: late November

Lying to the northwest of the Large Magellanic Cloud, the main stars of this small constellation form a narrow lozenge shape. 3rd magnitude Alpha Ret forms a challenging naked eye double with dim red 5th magnitude HIP19805 to its north. Zeta Ret is another challenging double of 5th magnitude yellow twins. Reticulum is lacking in bright deep sky objects.

Sculptor (the Sculptor)
Scl/Sculptoris; highest at midnight: early October

One of the sky's least conspicuous constellations, Sculptor hosts the South Galactic Pole. Northern Sculptor rises above the horizon and its whereabouts east of the bright star Fomalhaut is easy to locate. However, its faint stars are challenging to see even when the constellation is high in a dark southern hemisphere sky.

Positioned very close to the South Galactic Pole, the bright globular NGC 288 and the splendid nearly edge-on galaxy NGC 253 are under 2° apart and can be viewed together in the same low-power field. NGC 253 is bright, with an easily discernible nucleus. Through a 100mm telescope

considerable structure can be seen along the galaxy's arms, even though it is very foreshortened. NGC 253 lies in a field rich in foreground stars. NGC 55 is another bright, almost edge-on galaxy, with less obvious visual detail than can be seen in NGC 253.

◄ NGC 253, a nearly edge-on galaxy in Sculptor.

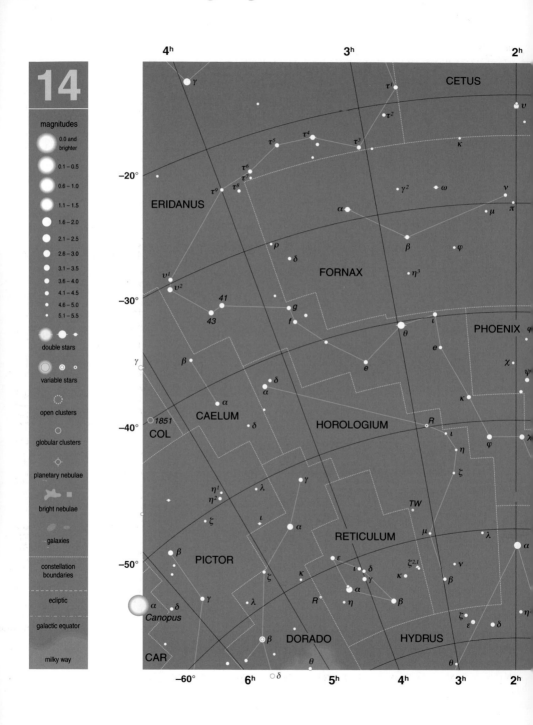

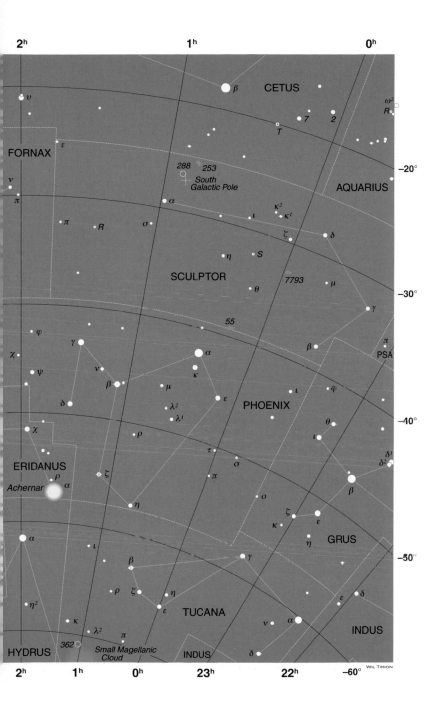

υ

CETUS β

ω²
R

FORNAX
ε
7 2

T

288 253
South
Galactic Pole

−20°

ν

π
α

AQUARIUS

π R σ

κ²
κ¹
ι

ζ δ

η S

SCULPTOR

θ 7793 μ

−30°

γ

55

φ

γ
α

π

χ

ν
κ

β
ι

PSA

ψ
μ

ε
ι φ

δ
λ²

PHOENIX

λ¹
θ

ρ
ι

−40°

τ
σ

δ¹

ERIDANUS

π

δ²

ρ
ζ
ο

β

Achernar
α

η

ζ
ε

α
κ
ε

η

GRUS −50°

ι
γ

β

ρ
ζ
η

ε
δ

η²
ε

TUCANA

ν
α

INDUS

κ
λ²

362
π

δ

HYDRUS
Small Magellanic
Cloud

INDUS

WIL TIRION

▶ # Chart 15

The sky's two brightest stars—Sirius in Canis Major and Canpous in Carina—blaze brilliantly in the eastern region of this chart. It is a chart of two distinct halves. The east, with its star-studded stretch of the Milky Way, is full of sublime magnificence, while the west makes far less of an immediate visual impact.

Caelum (the Chisel)
Cae/Caeli; highest at midnight: early December
Caelum is an uninteresting constellation—a narrow streak of sky containing just a handful of faint stars. The only object of interest is the double star Gamma Cae, comprising an orange mag 4.5 primary and a mag 8.1 companion, separable through a small telescope.

Canis Major (the Great Dog)
CMa/Canis Majoris; highest at midnight: early January
Canis Major is bright and easily recognizable. Alpha CMa (Sirius), the sky's brightest star, can be found by tracing the line of Orion's Belt southeast. Sirius is a famous binary with a faint white dwarf companion, Sirius B.

Some 4° south of Sirius lies the Little Beehive (M41), a bright open cluster just visible with the unaided eye as a 4th magnitude smudge. Its dozens of stars, visible through a small telescope, take up an area about as wide as the full Moon.

Canis Major's eastern sector is occupied by the Milky Way, and here can be found the bright open clusters of NGC 2360 and NGC 2362.

▶ The Tarantula Nebula in the Large Magellanic Cloud. Image by the Spitzer Space Telescope / NASA.

Columba (the Dove)

Col/Columbae; highest at midnight: mid-December

Situated between Sirius and Canopus, Columba is easy to locate, its main stars following a sinuous east-west path. The only object of interest is NGC 1851, a compact 7th magnitude globular cluster, resolvable through a 150mm telescope.

Dorado (the Goldfish)

Dor/Doradus; highest at midnight: early December

A short distance west and south of Canopus, the goldfish plunges diagonally across the sky, deep into southern climes, its stars mingling with the magnificent Large Magellanic Cloud (LMC). Dorado is easy to locate because of its proximity to the LMC. Most of the LMC is contained within southern Dorado. An irregular dwarf galaxy, the LMC contains around 10 billion stars. To the unaided eye it appears as a bright oval smudge, like a detached segment of the Milky Way. Through binoculars the LMC is a glorious sight, packed with detail. In the eastern LMC, the Tarantula Nebula (NGC 2070) can be seen with the unaided eye, forming a bright knot. Binoculars show it as a sizeable pinkish patch about the diameter of the full Moon. Its spidery outline is revealed through small telescopes, along with dozens of bright supergiants bustling at its center. In and around the LMC in southern Dorado are numerous splendid star clusters and nebulae, clearly visible through binoculars.

Pictor (the Easel)

Pic/Pictoris; highest at midnight: mid-December

Pictor is an inconspicuous constellation. Iota Pic, an easily resolvable double star of mag 5.6 and 6.4 components, is one of the few objects in Pictor worthy of the stargazer's attention.

Puppis (the Stern)

Pup/Puppis; highest at midnight: mid-January

Puppis, a large constellation lying east and south of Canis Major, spreads south across 40° of declination to the northern border of Carina. It is highest in the south on January nights in the US and is fully visible over the horizon. Puppis is crossed by the Milky Way and packed with bright open clusters.

Just 1° apart in the north of Puppis, M46 and M47 are a spectacular pair of open clusters, easily visible in the same binocular field. M93 is easy to find near Xi Pup, but it's smaller and less spectacular. NGC 2477, an open cluster west of Zeta Pup, can just be discerned with the unaided eye.

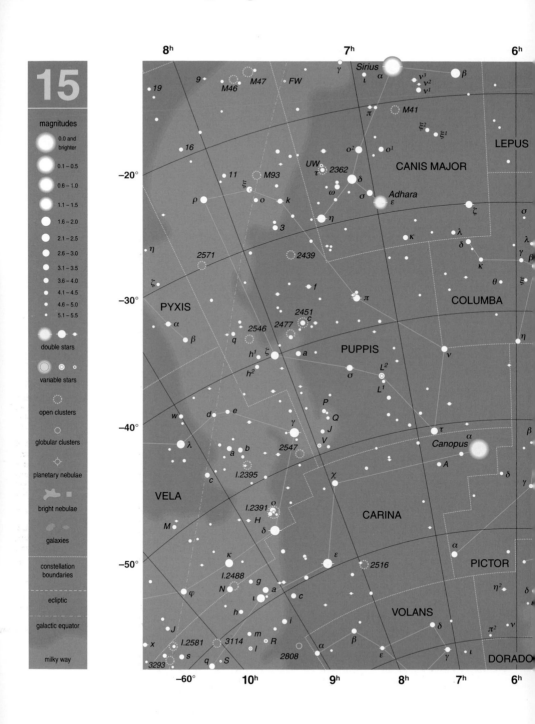

15

magnitudes

- 0.0 and brighter
- 0.1 – 0.5
- 0.6 – 1.0
- 1.1 – 1.5
- 1.6 – 2.0
- 2.1 – 2.5
- 2.6 – 3.0
- 3.1 – 3.5
- 3.6 – 4.0
- 4.1 – 4.5
- 4.6 – 5.0
- 5.1 – 5.5

double stars

variable stars

open clusters

globular clusters

planetary nebulae

bright nebulae

galaxies

constellation boundaries

ecliptic

galactic equator

milky way

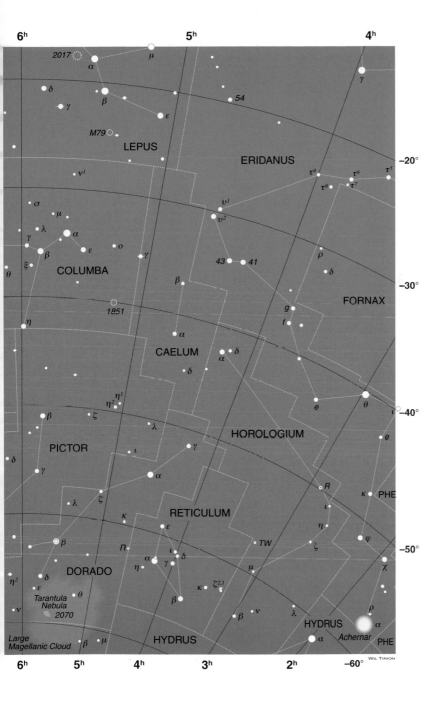

6ʰ   5ʰ   4ʰ

2017

μ

α

δ

γ   β   ε

54

γ

LEPUS

M79

ERIDANUS

−20°

ν¹

τ⁹   τ⁶   τ⁵

τ⁸   τ⁷

σ

μ

λ   υ¹

γ   α   υ²

β   ε   o

ξ   γ   **43   41**

ρ

θ   **COLUMBA**   δ

β   g

−30°

FORNAX

1851

f

α

CAELUM   δ

α

δ   e   θ   ι   −40°

η¹

η²   e

β   ζ

λ   **HOROLOGIUM**

PICTOR   ι   γ

δ   α   R

γ   ι   κ   **PHE**

λ   ζ   η   −50°

κ   **RETICULUM**   ζ

ε   φ

β   π   TW

DORADO   α   ι   δ   μ   χ

η   γ   ζ²˒¹

η²   δ   κ   ρ

ε   θ   β   β   ν   λ   **HYDRUS**   α

ν   *Tarantula*   *Nebula*   *2070*   α   *Achernar*   **PHE**

Large   *Magellanic Cloud*   β   μ   **HYDRUS**

6ʰ   5ʰ   4ʰ   3ʰ   2ʰ   −60°   Wıʟ Tıʀıon

▶ # Chart 16

A broad, magnificent section of the southern Milky Way, peppered with the brilliant stars of Vela, Carina, Crux, and Centaurus, makes an immense contrast with the relatively sparse region containing Hydra, Antlia, and Pyxis to the north.

Antlia (the Air Pump)

Ant/Antliae; highest at midnight: late February

A desolate, hard-to-locate southern constellation comprising a few scattered stars at the limit of naked eye visibility.

Carina (the Keel)

Car/Carinae; highest at midnight: early February

Carina is a large constellation spanning 40° of the southern circumpolar regions, from its brightest star Alpha Car (Canopus) in the west to its starry eastern realms. The outline made by Carina's brighter stars lies south of Vela; here can be found the False Cross, an asterism comprising stars of Carina and Vela which is sometimes mistaken for the constellation of Crux.

Eta Car (an unpredictable variable) is immersed within the Keyhole Nebula (NGC 3372), a large diffuse nebula visible with the unaided eye. Through binoculars and small telescopes it displays considerable structure, with delicate wisps streaming from discrete clumps of bright nebulosity, interwoven with dark lanes and peppered with dozens of stars.

NGC 3532 is a beautiful mass of stars in the 8th to 12th magnitude range, one of the finest open clusters. X Car, a 3rd magnitude yellow supergiant, lies far in the cosmic distance and can be seen on the cluster's eastern edge. Nearby, NGC 3293 is a smaller open cluster, remarkable for its mixture of bright blue and red stars. The Southern Pleiades (IC 2602) dazzles stargazers with its collection of bright stars spread around Theta Car, a number of which are visible with the unaided eye. NGC 2516 is another noteworthy open star cluster, visible without optical aid. Binoculars show the brightest of its 80 or so stars arranged in a striking cross shape.

Crux (the Southern Cross)

Cru/Crucis; highest at midnight: late March

Although the Southern Cross is the smallest constellation, it packs a lot of punch. Its four brightest stars, Alpha, Beta, Gamma, and Delta Cru, make a prominent cruciform pattern.

A small telescope resolves Alpha Cru into close twin components of mag 1.3 and 1.8. Another noteworthy double, Mu Cru, consists of a wide pair of mag 4 and 5.1.

▲ In one of the most spectacular parts of the sky, the Milky Way runs through Crux and Centaurus.

The Coalsack Nebula is a well-defined patch of darkness some 4° wide occupying the southeastern quadrant of Crux, intruding slightly into Musca and Centaurus. It is an isolated cloud of dust silhouetted against the Milky Way background. Visible as a misty spot with the unaided eye, just north of the Coalsack, the Jewel Box (NGC 4755) is a marvellous star cluster around Kappa Cru.

Pyxis (the Compass)
Pyx/Pyxidis; highest at midnight: early February
A small, faint, and insignificant constellation.

Vela (the Sails)
Vel/Velorum; highest at midnight: mid-February
Much of Vela is immersed in a bright section of the Milky Way. Its main outline—a broad oval consisting of a dozen naked eye stars—is easy to trace. Gamma Vel is its brightest star (there is no Alpha), an interesting multiple whose main components, blue stars of mag 1.8 and 4.3, are separable through a small telescope.

NGC 2547, is a rich treasure trove of around 80 stars. Vela's brightest cluster, IC 2391, can be seen without optical aid as a misty patch around Omicron Vel. It contains about 30 stars.

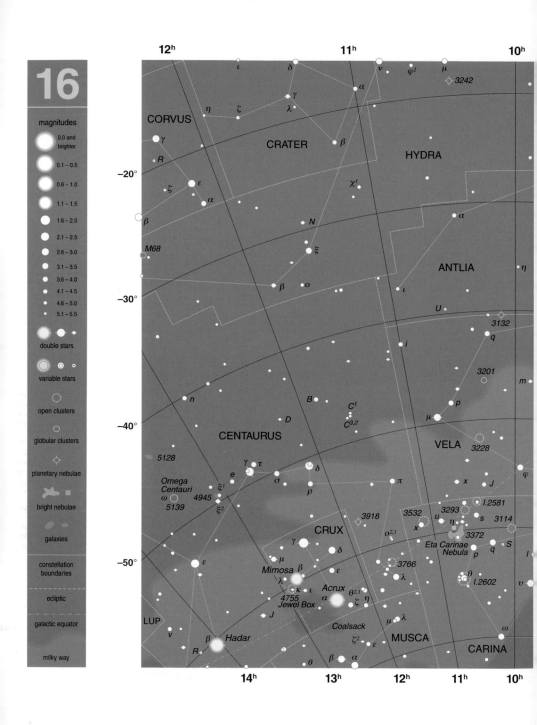

16

magnitudes

- 0.0 and brighter
- 0.1 – 0.5
- 0.6 – 1.0
- 1.1 – 1.5
- 1.6 – 2.0
- 2.1 – 2.5
- 2.6 – 3.0
- 3.1 – 3.5
- 3.6 – 4.0
- 4.1 – 4.5
- 4.6 – 5.0
- 5.1 – 5.5

double stars

variable stars

open clusters

globular clusters

planetary nebulae

bright nebulae

galaxies

constellation boundaries

ecliptic

galactic equator

milky way

12ʰ   11ʰ   10ʰ

CORVUS

CRATER

HYDRA

3242

ANTLIA

CENTAURUS

VELA

Omega Centauri

CRUX

Eta Carinae Nebula

LUP

MUSCA

CARINA

Mimosa

Acrux

Jewel Box

Coalsack

Hadar

14ʰ   13ʰ   12ʰ   11ʰ   10ʰ

−20°

−30°

−40°

−50°

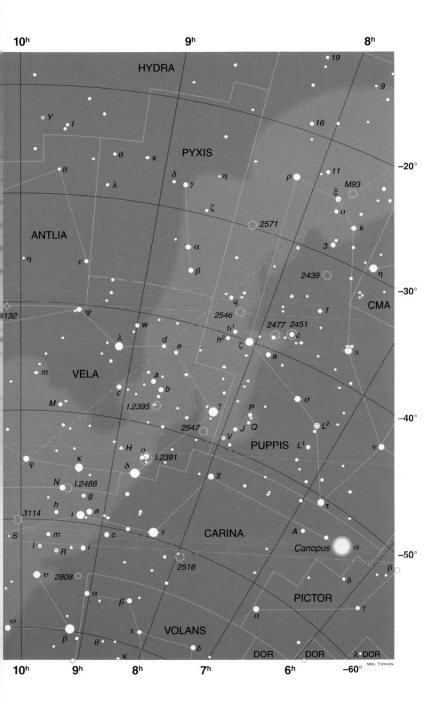

10ʰ · 9ʰ · 8ʰ

HYDRA

19

9

Y

l

16

PYXIS

θ

κ

θ

δ

γ

η

ρ

11

M93

ξ

−20°

λ

ζ

o

ANTLIA

2571

k

α

3

η

η

β

2439

−30°

ε

q

f

CMA

2546

132

ψ

w

h¹

2477 2451

c

λ

d

e

h²

ζ

π

VELA

a

a

b

σ

−40°

M

c

γ

P

φ

I.2395

J Q

L²

2547

V

ν

H

o

I.2391

PUPPIS

L¹

κ

δ

N

I.2488

χ

g

τ

3114

h

a

ι

S

A

CARINA

Canopus

α

−50°

m

c

ε

β

l

R

i

2516

δ

υ

2808

PICTOR

α

β

α

γ

ω

VOLANS

β

θ

ε

κ

δ

DOR · DOR · λ DOR

10ʰ · 9ʰ · 8ʰ · 7ʰ · 6ʰ · −60°

WIL TIRION

▶ # Chart 17

This region of the southern skies, from Scorpius in the east to Vela in the west, is dominated by the mighty constellation of Centaurus, which dips into the Milky Way between Crux and Lupus.

Centaurus (the Centaur)
Cen/Centauri; highest at midnight: mid-April

Centaurus' main stars are bright and easily traceable. More stars are visible with the unaided eye in Centaurus than any other constellation—at least 150 can be seen on a clear dark evening. Binoculars bring some astounding Centauran star fields into view, particularly along the bright section of the Milky Way in the south.

Centaurus' brightest stars, the yellow Alpha Cen (Rigil Kent) and blue Beta Cen (Hadar), form a lovely pair. Just 4.4 light years away, Alpha Cen is the nearest bright star. A small telescope reveals that it is a double star, comprising twin components of mag 0 and 1.35. The system also includes a dim (11th magnitude) red dwarf called Proxima Cen, 2° southwest of Alpha Cen and visible only in the same field of view through a large telescope using a low-magnification, wide-field eyepiece. Proxima is currently nearer to us by just 0.2 light years, making it the Sun's closest stellar neighbor. Beta Cen has components of 0.6 and 3.9, separable in a 150mm telescope, difficult because of the primary's brightness.

West of 2nd magnitude Zeta Cen, a fuzzy 3rd magnitude star can be discerned with the unaided eye. This is Omega Centauri (NGC 5139), the sky's biggest and brightest globular cluster—a sphere of more than a million stars occupying an area equal to the full Moon. It is majestic

▶ Omega Centauri, a mighty globular cluster containing more than a million stars.

through any instrument, but fantastic detail can be discerned through any telescope larger than 150mm. East of Zeta Cen lies NGC 5460, a lovely open cluster on the verge of naked eye visibility, although its 40 or so stars are all fainter than 8th magnitude.

Circinus (the Pair of Compasses)
Cir/Circini; highest at midnight: early May

Although the small constellation of Circinus is easy to locate east of Rigil Kent, its stars are faint. Circinus spreads over a bright part of the Milky Way, so it is pleasant to scan through binoculars. The 3rd magnitude Alpha Cir has a faint 8th magnitude partner, an easy telescopic sight. Gamma Cir is a lovely pair of 5th magnitude blue and yellow stars, more difficult to split, requiring at least a 150mm telescope.

Lupus (the Wolf)
Lup/Lupi; highest at midnight: mid-May

Lupus lies along the northern edge of the Milky Way, east of Centaurus across to the western border of Scorpius. U.S. stargazers get their best view of the brightest stars of Lupus on fresh spring mornings.

Eta Lup is a nice double with a blue mag 3.4 primary and a yellow mag 7.8 companion, resolvable through a small telescope. Kappa Lup is a wide double, mags 3.9 and 5.7. The multiple Mu Lup consists of an easily splittable blue mag 4.3 primary and a mag 6.9 partner. Xi Lup is a pair of blue 5th mag stars, delightful through a small telescope.

Nestled within the bright Milky Way starfields southwest of Zeta Lup, the open cluster NGC 5822 appears as a Moon-sized patch through binoculars. Small telescopes resolve its 150 or so loosely assembled stars, among which can be discerned a winding stellar chain. NGC 5986 is the brightest globular cluster within Lupus, small but resolvable to its bright core using a 200mm telescope.

Norma (the Level)
Nor/Normae; highest at midnight: late May

Norma is a small constellation made up of a few faint stars set against the Milky Way. Epsilon Nor is an easily splittable double of mag 4.4 and 6.1. Another easily resolved double is Iota 1 Nor, of mag 4.6 and 8.8. To its east lies the open cluster NGC 6087, many of its stars resolvable through binoculars. Another fine open cluster of about equivalent brightness is NGC 6067. This contains around 100 stars of the 10th magnitude and fainter.

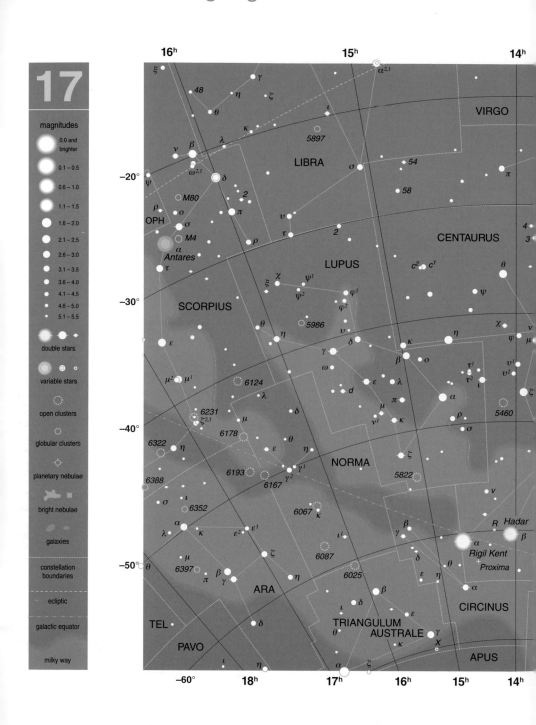

17

magnitudes

- 0.0 and brighter
- 0.1 – 0.5
- 0.6 – 1.0
- 1.1 – 1.5
- 1.6 – 2.0
- 2.1 – 2.5
- 2.6 – 3.0
- 3.1 – 3.5
- 3.6 – 4.0
- 4.1 – 4.5
- 4.6 – 5.0
- 5.1 – 5.5

double stars

variable stars

open clusters

globular clusters

planetary nebulae

bright nebulae

galaxies

constellation boundaries

ecliptic

galactic equator

milky way

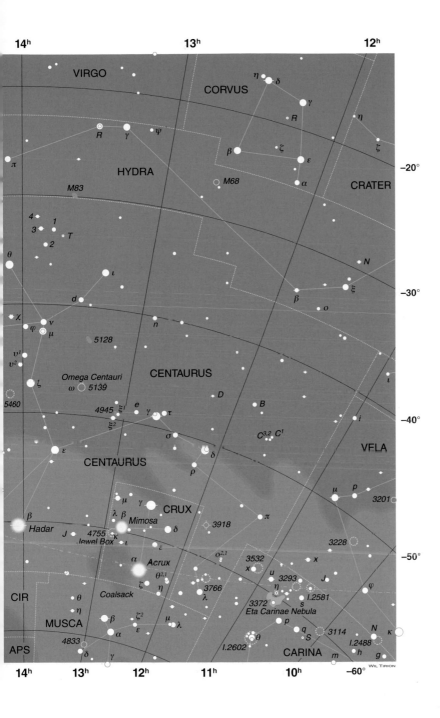

WIL TIRION

▶ # Chart 18

A grand section of the southern Milky Way—representing the enormous central bulge of our home galaxy—broadens to its widest in Sagittarius. Our view of the galactic center is obscured by dark bands of dust and gas. Here and there it is illuminated by wonderful nebulae.

Ara (the Altar)

Ara/Arae; highest at midnight: mid-June

Ara is a small constellation made up of a collection of 2nd and 3rd magnitude stars set against the Milky Way. The open cluster NGC 6193, visible with the unaided eye, contains about 30 stars; its bright central star is an easily separable double of mag 5.6 and 6.8. Ara contains three globular clusters—NGC 6352, NGC 6397, and NGC 6362.

Corona Australis (the Southern Crown)

CrA/Coronae Australis; highest at midnight: early July

Corona Borealis is a little constellation whose brightest stars form an arc. Kappa 2 and Kappa 1 CrA form a lovely blue double comprising a mag 5.9 primary and mag 6.6 partner, easily separated with a small telescope. NGC 6541 is a rare example of a globular within the Milky Way. It is visible through binoculars as a 7th magnitude smudge with a bright core.

Pavo (the Peacock)

Pav/Pavonis; highest at midnight: mid-July

Pavo is a broad southern circumpolar constellation made up of sub-3rd magnitude stars, with the exception of 1st magnitude Alpha Pav.

Just visible with the unaided eye, NGC 6752 is a bright, compact, globular cluster which begins to be resolved through an 80mm telescope.

Sagittarius (the Archer)

Sag/Sagittarii; highest at midnight: early July

Sagittarius is crossed by a broad swathe of the Milky Way, and beyond it, the center of our galaxy, near the constellation's western border.

No fewer than 15 Messier objects lie within Sagittarius, most of which appear against the backdrop of the Milky Way. The Lagoon Nebula (M8) is a superb diffuse nebula, visible without optical aid as a Moon-sized glow.

▶ The Lagoon Nebula and the Trifid Nebula—two impressive deep sky sights in Sagittarius.

Telescopes show a dark band within M8. NGC 6530, an open cluster, shines within the nebula's eastern glow. One degree to its north, the Trifid Nebula (M20) is also visible with the unaided eye, and to its immediate north lies M21, an open cluster. M20 is smaller than M8, but it is just as spectacular. Three dark lanes visible within the nebula give it its name. M8, NGC 6530, M20, and M21 can all be seen together in the same low-power, wide-angle field.

Scorpius (the Scorpion)
Sco/Scorpii; highest at midnight: early June

The collection of Scorpius' bright stars near Alpha Sco (Antares), in southern summer skies, are a familiar sight to northern stargazers. The faint blue companion to the orange supergiant Antares may be glimpsed through a small instrument, though the brilliance of Antares hinders its visibility. West of Antares, the globular cluster M4 can be seen with the unaided eye; a 100mm telescope will resolve it.

Two magnificent open clusters are visible without optical aid. Set like a large garnet upon the eastern wing of the diamond brooch of M6 is the orange giant BM Sco; 3° southeast of M6, the larger and brighter M7 is dazzling. Binoculars reveal many of the brighter stars. Through a telescope at low magnifications it is impressive, its central stars forming a distinct H shape.

Triangulum Australe (the Southern Triangle)
TrA/Trianguli Australis; highest at midnight: late May

Triangulum Australe is a small constellation immersed in a bright section of the Milky Way. NGC 6025 is a bright open cluster comprising around 60 stars set against a lovely galactic starfield.

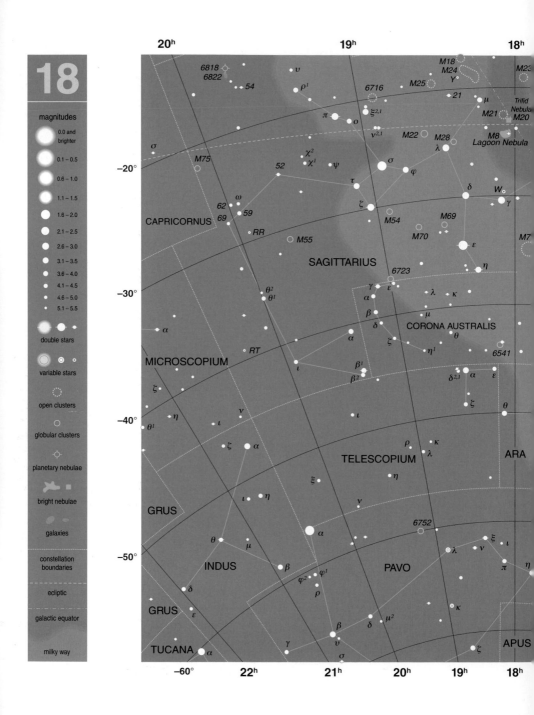

18

magnitudes

- 0.0 and brighter
- 0.1 – 0.5
- 0.6 – 1.0
- 1.1 – 1.5
- 1.6 – 2.0
- 2.1 – 2.5
- 2.6 – 3.0
- 3.1 – 3.5
- 3.6 – 4.0
- 4.1 – 4.5
- 4.6 – 5.0
- 5.1 – 5.5

double stars

variable stars

open clusters

globular clusters

planetary nebulae

bright nebulae

galaxies

constellation boundaries

ecliptic

galactic equator

milky way

20ʰ   19ʰ   18ʰ

6818
6822   54   υ
ρ¹
π   ξ²,¹   6716   M25
o   M18
M24   Y°
21   μ   M23
ν²,¹   M22   M28   Trifid
Nebula
M21   M20
σ   M75   χ²   χ¹   ψ   λ   M8
Lagoon Nebula
52   τ   σ   φ   δ   W
62   ω   γ
69   59   ζ   M54   M69   M7
CAPRICORNUS   RR   M55   M70   ε
θ²   SAGITTARIUS   6723   η
θ¹   γ   ε   λ   κ
α   α   μ
β   δ   CORONA AUSTRALIS
α   ζ   θ
RT   η¹   6541
ι   β¹   δ²,¹   α   ε
MICROSCOPIUM   β²
ζ   ι   ζ
η   ν   θ
θ¹   ι   ρ   κ   ARA
ζ   α   λ
TELESCOPIUM
ξ   η
GRUS   ι   η   γ
6752
θ   α   ξ   ι
INDUS   μ   λ   ν
β   PAVO   π   η
φ²   φ¹   κ
δ   ρ
GRUS   β   μ²
ε   δ
γ   ν   ζ   APUS
TUCANA   α   σ

−20°

−30°

−40°

−50°

−60°

22ʰ   21ʰ   20ʰ   19ʰ   18ʰ

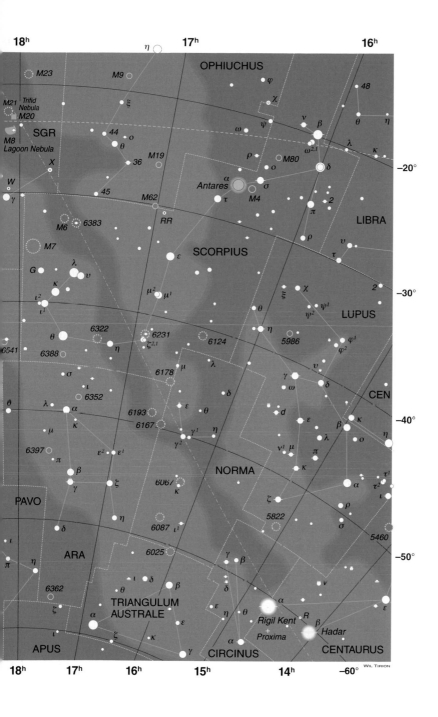

18ʰ   17ʰ   16ʰ

η

M23

M9

φ

48

M21   Trifid
Nebula
M20

ξ

χ

ν   β

θ   η

OPHIUCHUS

ψ

ω

λ

κ

SGR

44   ο

ρ

ω²·¹

δ   −20°

θ

36

M19

ο   M80

M8
Lagoon Nebula

X

α
Antares   σ

2

W

45

M62

τ   M4

π

LIBRA

γ

RR

ρ

υ

M6   6383

τ

M7

ε

SCORPIUS

2   −30°

G

λ
υ

ξ   χ

κ

μ²   μ¹

ψ¹

LUPUS

ι²
ι¹

θ

ψ²

φ¹

θ   6322

6231

6124

5986

φ²

6541

η

6388

ζ²·¹

6178

μ

λ

ν

CEN   −40°

σ

δ

γ

ω   δ

ι   6352

ε

θ

d

η

θ   λ

α

6193

ε

β   κ

μ

κ

6167

γ¹   η

λ

π

τ¹

6397

π

γ²

ν¹   μ

τ²

β

ε²   ε¹

κ

ι

NORMA

ρ

γ

ζ

6067

α

κ

ζ

σ

5822

η

6087

ι¹

5460   −50°

δ

6025

γ   β

PAVO

ι

ARA

η

γ

δ

ν

π

6362

θ   ι   δ

β

δ

ε

ζ

α

ε

η   θ

α
Rigil Kent

R   β

ε

TRIANGULUM
AUSTRALE

γ

+
Proxima

Hadar

APUS

ζ

κ

α

CENTAURUS

ι

γ

CIRCINUS

18ʰ   17ʰ   16ʰ   15ʰ   14ʰ   −60°

WIL TIRION

▶ # Chart 19

An assortment of medium-sized constellations is arrayed across the region depicted, some of which are obscure, others easy to identify. Devoid of the Milky Way and with a dearth of bright deep-sky showpieces, this is one of the least glamorous stretches of the entire heavens.

Grus (the Crane)
Gru/Grucis; highest at midnight: late August
South of the bright star Fomalhaut, Grus is easy to locate. Its brightest stars form an easily traceable cross shape, similar to its northern cousin, Cygnus. Both Delta Gru and Mu Gru are wide doubles separable with a keen naked eye. Binoculars will reveal the dual nature of Pi Gru, a red variable of 5-6th magnitude with a white 5th mag line-of-sight companion.

Indus (the Native American Indian)
Ind/Indi; highest at midnight: mid-August
Squeezed between Grus and Pavo, the south circumpolar constellation of Indus comprises several 3rd and 4th magnitude stars with a scattering of fainter stars further to its south. Indus' only object of visual impact is Theta Ind, a double comprising a pair of white mag 4.5 and 7 stars.

Microscopium (the Microscope)
Mic/Microscopii; highest at midnight: early August
Microscopium is one of the most obscure constellations, made up of a scattered bunch of stars at the limit of naked eye visibility, encapsulated within a small box due south of Capricornus.

Phoenix (the Firebird)
Phe/Phoenecis; highest at midnight: early October
Located immediately north and northwest of Achernar in Eridanus, Phoenix is easy to find, its distorted W zigzag pattern made up of its five brightest stars. Beta Phe, the central star in the W, is a close double made of twin 4th magnitude red giants, separable through an 80mm telescope.

Piscis Austrinus (the Southern Fishes)
PsA/Piscis Austrini; highest at midnight: late August
Piscis Austrinus lies on a cold rectangular celestial fishmonger's slab, a

rather unspectacular stellar array, save for bright blue Alpha PsA (Fomalhaut). Shining at 1st magnitude, Fomalhaut dominates its celestial surroundings. It can be seen above the southern horizon during darkening Fall evenings.

Beta PsA is a wide double, resolvable through binoculars, with components of mag 4.5 and 7.5. Delta and Gamma PsA, immediately south of Fomalhaut, form a wide naked eye double. 4th magnitude Delta PsA has a faint 10th magnitude companion, which is visible through an 80mm telescope. Gamma PsA has components of mag 4.5 and 8.5, visible through a small telescope. Eta PsA is a very close double of white mag 5.4 and 6.6 stars, which is resolvable through a 100mm telescope.

Telescopium (the Telescope)

Tel/Telescopium; highest at midnight: early July

Faint and uninteresting, Telescopium covers a rectangle of sky south of Corona Australis. Its boundaries enclose few objects of telescopic interest. Delta Tel is a wide optical double of mag 4.9 and 5.1 near Alpha Tel, separable with a really keen unaided eye.

◀ Fomalhaut nudges above the southern horizon looking south on Fall evenings. It can be found by following the line of the western stars of the Square of Pegasus.

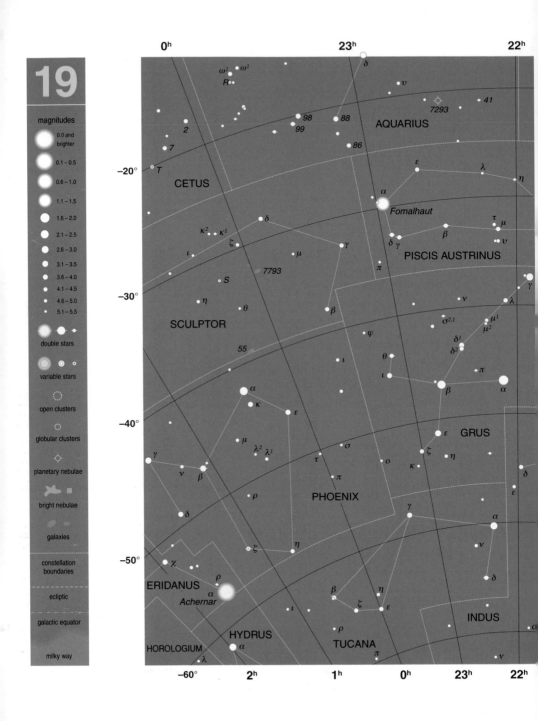

19

magnitudes

- 0.0 and brighter
- 0.1 – 0.5
- 0.6 – 1.0
- 1.1 – 1.5
- 1.6 – 2.0
- 2.1 – 2.5
- 2.6 – 3.0
- 3.1 – 3.5
- 3.6 – 4.0
- 4.1 – 4.5
- 4.6 – 5.0
- 5.1 – 5.5

double stars

variable stars

open clusters

globular clusters

planetary nebulae

bright nebulae

galaxies

constellation boundaries

ecliptic

galactic equator

milky way

0ʰ   23ʰ   22ʰ

ω² ω¹
R

98   88
99
86

AQUARIUS

7293   41

–20°   ε   λ   η
T   α
CETUS   Fomalhaut   τ μ
β   ν
κ² κ¹   δ   δ γ
ξ   PISCIS AUSTRINUS
ι   μ   γ
7793   π
S
η   θ   β   ν
–30°   γ   λ
σ²·¹
SCULPTOR   μ¹
φ   μ²
55   δ¹
θ   δ²
ι   π
ι   β   α
α
κ   ε
–40°   ε   GRUS
μ   σ   ζ   η
γ   λ² λ¹   τ   o   κ
ν   β   π   PHOENIX
ρ   γ   α
δ   ν
ζ   η   δ
–50°   ε
χ   β   η
ERIDANUS   ρ   ζ   ε   INDUS
α
Achernar   ρ
HYDRUS   α   TUCANA
HOROLOGIUM   π
λ   ν

–60°   2ʰ   1ʰ   0ʰ   23ʰ   22ʰ

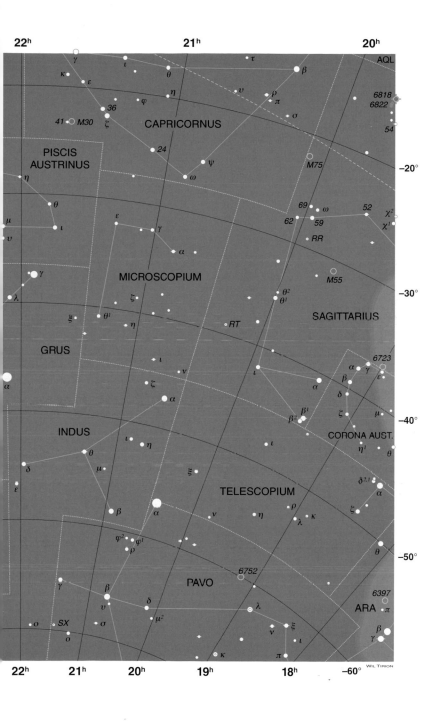

γ
ι
κ
ε
θ
τ
AQL
η
υ
ρ
β
φ
π
6818
6822
36
σ
41 M30 ζ
54
CAPRICORNUS
PISCIS
AUSTRINUS
24
−20°
ψ
η
ω
M75
θ
ε
69 ω 52 χ²
μ
ι
γ
62 59 χ¹
ν
α
RR
γ
MICROSCOPIUM
M55
−30°
λ
ζ
θ²
θ¹
ξ
θ¹
η
RT
SAGITTARIUS
GRUS
ι
ν
α 6723
ζ
ι
α γ ε
α
β
INDUS α
δ
β¹
ι
η
β² μ
ζ
−40°
CORONA AUST.
θ
μ
ξ
ι
η¹
θ
δ
ε
δ²ᐟ¹
TELESCOPIUM
α
β α
ν
η ρ
ζ
κ
λ
φ² φ¹
θ
ρ
−50°
6752
PAVO
6397
γ
β
ARA π
υ
δ
λ
o SX
μ²
ξ
β
o σ
ν
ι
γ
κ
π

WIL TIRION

▶ # Chart 20

There are two distinct halves to this chart of the southern circumpolar region, more dramatically contrasted than that shown around the north celestial pole. On one side lies the glow of the Milky Way, bubbling with brilliant stars and nebulae, while the other side is devoid of bright stars, beyond which shine the two patches of the Large and Small Magellanic Clouds.

Apus (the Bird of Paradise)
Aps/Apodis; highest at midnight: late May
Apus is a faint circumpolar constellation. About the only object of interest to the stargazer is Delta Aps, a binocular double comprising a lovely red mag 4.7 primary and orange mag 5.3 companion.

Chamaeleon (the Chamaeleon)
Cha/Chamaeleontis; highest at midnight: early March
Another unimpressive southern circumpolar constellation. Its stars are faint, all of the 4th magnitude and below, the brightest arranged in an elongated lozenge. A lovely color contrast is to be seen in Delta Cha, a wide optical double consisting of a blue mag 4.4 star and an orange mag 5.5 star. Just 1.5° to its west lies NGC 3195, the sky's most southerly planetary nebula.

Hydrus (the Lesser Water Snake)
Hyi/Hydri; highest at midnight: late October
Hydrus can easily be located, since its three bright main stars form a triangle between the Large and Small Magellanic Clouds. Pi Hyi forms a nice pair of orange 5th magnitude twins, separable through binoculars.

Mensa (the Table Mountain)
Men/Mensae; highest at midnight: mid-December
Mensa contains fewer bright stars than any other constellation—all are 5th magnitude and below.

Musca (the Fly)
Mus/Muscae; highest at midnight: late March
Musca is an easily identifiable collection of fairly bright stars spanning a bright section of the Milky Way.

Half a degree north of Delta Mus lies the Southern Butterfly (NGC 4833), visible through binoculars as a fuzzy spot and resolvable using a 150mm telescope. Several faint loops of unresolved stars emanate from either side, giving it the appearance of a butterfly—compare with the Butterfly Cluster (M6) in Scorpius, as the two are often high in the sky at the same time.

Octans (the Octant)
Oct/Octantis; highest at midnight: early August
Octans, an unimpressive constellation whose main stars are of the 3rd and 4th magnitude, is home to the south celestial pole. Mag 5.4 Sigma Oct is the closest naked eye star to the pole, just one degree from it—an angular distance slightly larger than that between Polaris and the north celestial pole.

Tucana (the Toucan)
Tuc/Tucanae; highest at midnight: mid-September
Tucana's main stars are difficult to trace with the unaided eye, but its location can be identified easily, since it encompasses the Small Magellanic Cloud (SMC). Binoculars show that Beta Tuc comprises beautiful twin blue 4th magnitude stars. Another noteworthy double, Kappa Tuc, consists of a mag 5.1 primary and a mag 7.3 companion, divisible through a small telescope.

The SMC, a small irregular galaxy 200,000 light years from our own Milky Way, appears as a glowing oval patch. It is a splendid object to scan through binoculars. On its southern edge lie two bright globular clusters; 47 Tuc (NGC 104) is a large globular, visible without optical aid, and glorious through binoculars; NGC 362 is a bright globular on the border of naked eye visibility.

Volans (the Flying Fish)
Vol/Volantis; highest at midnight: mid-January
A small constellation between the Milky Way and the LMC, Volans is made up of a "little dipper" pattern of 4th magnitude stars. Gamma Vol is a double with a yellow mag 3.8 primary and a white mag 5.7 companion. Epsilon Vol is another telescopic double, comprising a blue mag 4.3 primary and a mag 7.3 companion.

◀ Deep sky delights in the Small Magellanic Cloud.

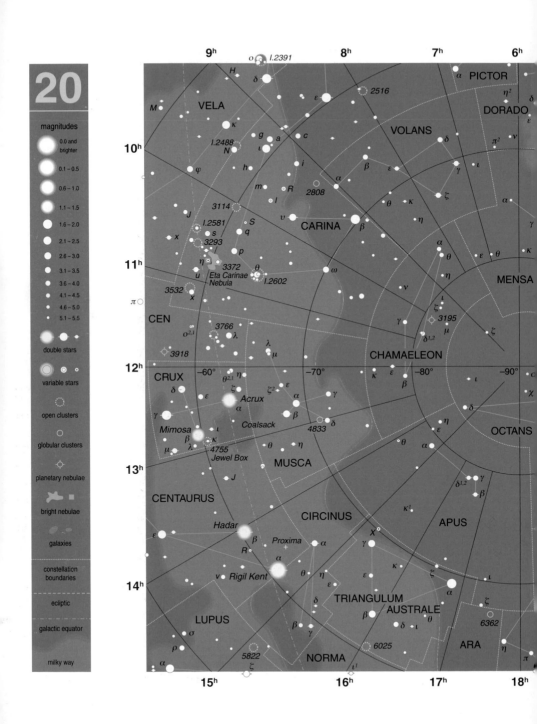

20

magnitudes

- 0.0 and brighter
- 0.1 – 0.5
- 0.6 – 1.0
- 1.1 – 1.5
- 1.6 – 2.0
- 2.1 – 2.5
- 2.6 – 3.0
- 3.1 – 3.5
- 3.6 – 4.0
- 4.1 – 4.5
- 4.6 – 5.0
- 5.1 – 5.5

double stars

variable stars

open clusters

globular clusters

planetary nebulae

bright nebulae

galaxies

constellation boundaries

ecliptic

galactic equator

milky way

9ʰ   8ʰ   7ʰ   6ʰ

o I.2391

H
δ
ε
2516   α PICTOR
η²   δ
M   VELA   DORADO
κ   VOLANS   ε
10ʰ   g a c   δ
I.2488   ι   β ε   π² ν
N   h   γ   ι
φ   i   α
m   R   2808   θ κ   ζ   α
3114   l   η   γ
J   I.2581   υ   CARINA   β   α
x   S   θ   ε θ   κ
s   q   α   η
3293   p   ω   ν   MENSA
η   3372   θ   ζ   3195
u   Eta Carinae   I.2602   γ   μ   ζ
π   3532   Nebula   δ¹·²
CEN   x   3766   CHAMAELEON   -90°
O²·¹   λ   κ ε   -80°   ι   χ
3918   λ   -70°   β
12ʰ   CRUX   -60°   η   μ   ι
δ   θ²·¹   ζ² ε   δ   OCTANS
ε   ζ   α   γ   η
γ   Acrux   β   δ   ε   α
α   Coalsack   4833   θ
Mimosa   ι   θ   η   α
β κ   MUSCA
μ   λ   4755
13ʰ   Jewel Box   J   δ¹·² γ
CENTAURUS   κ¹   β
CIRCINUS   APUS
Hadar   X
ε   β   Proxima   α   γ   κ¹
R   ε   ζ   ι
14ʰ   ν Rigil Kent   θ η   α   ζ
ε   δ   TRIANGULUM   6362
LUPUS   β   AUSTRALE   δ ι   θ
β   γ   6025   ARA   η
σ   6025   π
ρ   5822   NORMA   ι¹
α   ζ

15ʰ   16ʰ   17ʰ   18ʰ

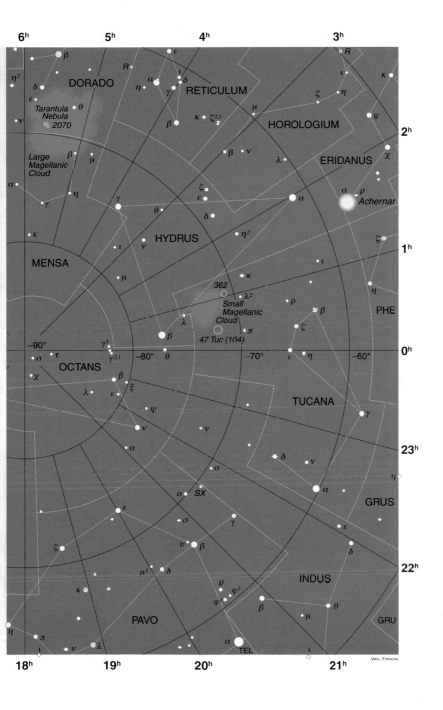

WIL TIRION

Glossary

Asteroid
A large chunk of rock orbiting the Sun, from a few yards to a few hundred miles in diameter. Also called a minor planet.

Asteroid belt
A zone of the Solar System containing a large number of asteroids. The main asteroid belt lies between the orbits of Mars and Jupiter.

Astronomical Unit (AU)
The average distance from the Earth to the Sun—around 93 million miles.

Astronomy
The science-based study of celestial objects and phenomena.

Aurora
The light show produced when energetic particles from the Sun get trapped in the Earth's magnetic field and collide with molecules in the Earth's upper atmosphere. The Aurora Borealis takes place in the northern hemisphere; the Aurora Australis takes place in the southern hemisphere.

Big Bang
The colossal explosion at the beginning of time that created the universe and everything in it.

Binoculars
An optical instrument consisting of a pair of small parallel refracting telescopes, allowing both eyes to view simultaneously.

Black hole
A relatively small region of collapsed space-time with such a huge gravity field that nothing—not even light itself—can escape.

CCD
A charge-coupled device, a light-sensitive electronic chip used in astrophotography.

Celestial sphere
From our perspective on the Earth's surface, the stars appear attached to the inside of a vast sphere. The Earth's poles point directly to the celestial poles, and the celestial sphere appears to rotate around us as the Earth spins on its axis.

Comet
A city-sized chunk of ice and rock which heats up on entering the inner Solar System, emitting gas and dust which forms a coma, and perhaps developing a long tail of gas and dust.

Constellation
A precisely defined region of the sky created to enhance our familiarity with the heavens. There are 88 recognized constellations, some of which date back to antiquity.

Core
The central region of a star or large planet —usually very hot and under extreme pressure. Many minor planets are chips from larger objects and may not have a well-defined core.

Double star
A pair of stars which appear close together in the sky. Some are binary systems orbiting each other; others, produced by line-of-sight perspective, are known as optical doubles. Systems of three or more stars are called multiple stars.

Eclipse
A phenomenon caused when a celestial object passes in front of, or through the shadow of, another celestial object. The Moon sometimes eclipses the Sun; the Moon itself is sometimes eclipsed by the shadow of the Earth.

Exoplanet
A planet in orbit around a distant star. Hundreds are currently known.

Eyepiece
A lens inserted into a telescope which magnifies light and focuses it into the eye.

Fireball
A very bright meteor caused by a large meteoroid's passage through the Earth's atmosphere.

Galaxy
A large-scale agglomeration of matter, in which may be contained 100 billion stars or more, held together by gravity and usually

centered around a massive hub of stars. Galaxies come in a variety of forms, some spiral, some elliptical, some irregular in shape.

Gas giant
A very large planet which is composed largely of gas—mainly hydrogen and helium. Jupiter, Saturn, Uranus, and Neptune are the Solar System's gas giants. They have no solid surface.

Globular cluster
A collection of hundreds of thousands of individual stars, all held together in a vast sphere by their mutual gravity.

Light year
The distance travelled by light in one year. At a velocity of 186,000 miles per second, light travels around 6 trillion miles in a year.

Magnitude
The perceived brightness of a celestial object is called its apparent magnitude. Its real brightness, called absolute magnitude, takes into account the object's distance.

Meteor
A flash of light caused when a meteoroid burns up on entering the upper atmosphere.

Meteoroid
A small lump of rock in space. If it survives all the way down to the Earth's surface, it is called a meteorite.

Milky Way
The name given to our home galaxy. The distant stars in the galactic plane can be seen in a misty band encircling the sky.

Moon
The Moon is the Earth's only natural satellite. Satellites around other planets are also referred to as moons.

Nebula
A cloud of interstellar dust and gas. It may shine by reflecting the light from nearby stars, or by emitting its own light. Dark nebulae appear silhouetted against a brighter background. Old stars can be surrounded by planetary nebulae, well-defined shells of puffed-off gas.

Planet
A large non-stellar object in orbit around a star. The Sun has nine major planets.

Reflector
A telescope that uses a large mirror to collect and focus light.

Refractor
A telescope that uses a large lens to collect and focus light.

Satellite
Any object in orbit around a larger body. Most planets have satellites.

Solar System
Our cosmic backyard, containing the Sun and everything that orbits the Sun, including the planets and their satellites, asteroids, and comets.

Star
A huge ball of incandescent gas shining by nuclear fusion. The Sun is a star.

Stargazer
An intelligent person who enjoys viewing the night skies, and often contemplates life, the universe, and everything, while doing so.

Sunspot
A slightly cooler region on the surface of the Sun which appears dark against the brighter background.

Supernova
The catastrophic explosion of a giant star at the end of its life.

Telescope
An optical instrument that produces magnified images of distant objects by using lenses and/or mirrors to collect and focus light. Different kinds of telescopes also collect other electromagnetic wavelengths, such as radio waves or infrared radiation, to produce images of phenomena invisible to the eye.

Variable star
A star whose apparent brightness fluctuates over time, either through being eclipsed by an orbiting companion or through changes in its size and/or the level of its light output.

Need to know more?

Books are no substitute for actually meeting, enthusing with, and learning from fellow lovers of the night skies, and joining a local astronomy club or society is a great way of sharing your stargazing passion. Most large cities, towns, and regions across the globe have their own astro-societies.

Resources

The Astronomical League
National Office Manager
9201 Ward Parkway Suite #100
Kansas City, MO 64114
www.astroleague.org
An online directory of local astronomy societies.

Association of Lunar and Planetary Observers (ALPO)
www.lpl.arizona.edu/alpo

American Lunar Society
www.otterdad.dynip.com/als/
A group dedicated to the continued study of the earth's moon.

Astronomy Magazine
www.astronomy.com
A monthly magazine covering wide ranging astronomical topics.

International Dark Sky Association (IDA)
3225 N. First Avenue
Tucson, Arizona 85719-2103 USA
www.darksky.org
An association that aims to build awareness of light pollution and provide solutions to the problem.

Smithsonian Institution's National Air and Space Museum
www.nasm.si.edu

National Aeronautics and Space Administration (NASA)
www.nasa.gov

Websites

Astronomy Picture of the Day
http://antwrp.gsfc.nasa.gov/apod/astropix.html
Each day, a different image of the universe is featured, along with a brief explination written by a professional astronomer. Addictive!

The Astronomy Net
www.astronomy.net
An online source for numerous astronomy resources that includes discussion forums, astro images, and constellation guides.

American Association of Variable Star Observers
www.aavso.org

Astronomical software

Starry Night
www.starrynight.com
A range of excellent planetarium programs aimed at abilities.

Redshift 5
www.redshift.de/
A super planetarium program with many useful features for planning your observations.

Books

Dunlop, Storm, *Atlas of the Night Sky* (HarperCollins, 2005)

Grego, Peter, *Moon Observer's Guide* (Firefly, 2004)

Grego, Peter, *Moonwatch* (Firefly, 2004)

Heifetz, Milton D. & Tirion, Wil, *A Walk Through the Heavens* (Cambridge, 2004)

Inglis, Mike, *Astronomy of the Milky Way* (Springer, 2004)

Levy, David, *Skywatching* (HarperCollins, 1995)

Menzel, Donald, *A Field Guide to the Stars and Planets* (Houghton Mifflin, 1975)

Moore, Patrick, *The Amateur Astronomer* (Cambridge, 1990)

Phillips, T. E. R., and Stevenson, W. H., *Splendor of the Heavens* (Hutchinson, 1923)

Ridpath, Ian, *Collins Gem—Stars* (HarperCollins, 2004)

Ridpath, Ian, *The Illustrated Encyclopedia of the Universe* (Watson-Guptill, 2001)

Ridpath, Ian, *Norton's Star Atlas* (Pi, 2003)

Ridpath, Ian, & Tirion, Wil, Collins *Pocket Guide to Stars and Planets* (HarperCollins, 1994)

Rukl, Antonin & Rackham, Thomas W., *Atlas of the Moon* (Kalmbach, 1992)

Rukl, Antonin, *Constellation Handbook* (Sterling, 2003)

Webb, Thomas, *Celestial Objects for Common Telescopes* Longmans, Green & Co, 1873)

Picture credits

Paul Sutherland (top) 9, (btt right) 55; Peter Vasey (btt) 9, 90, 110–1, 138–9, 151, 167; Anthony Ayiomamitis 11, (btt) 58, (btt) 59, (btt) 64; NASA 60; NASA/JPL (top) 75, 76; SwRI/NASA (btt) 75; Galileo/NASA (btt) 78, (btt) 79; NASA/STScI 17, 69, 72, (btt) 74, (top) 78, (top) 79, 183; NASA/STScI/AURA 89, 91, 94–5, 96–7, 99; Richard Bailey (top left) 52, 56, (top) 58; Dusko Novakovic (btt right) 52, 53; Colin Ebdon (btt right) 53; Ian Brantlingham 54; Mike Goodall 57, 65; Kev Smith (top) 59; Lee Macdonald (btt) 62; Dave Tyler 70; Cassini Imaging Team (top) 73; Nik Szymanek 88, 107, 115, 127, 150, 159, 170, 175; Paul Jenkins 93, (btt) 119, 146; Ian Robson (top) 119; Peter Grego/Wil Tirion 103; all remaining pictures copyright Peter Grego.

Index

ALPO (Association of Lunar and Planetary Observers) 45
Annual meteor showers 54
Asteroids 76, 78–9
Astronomical Unit 14
Astronomy, history of 10–17
Atmospheric effects 52
Aurora australis 53
Aurora borealis 53
Averted vision 27

BAA (British Astronomical Association) 45
Barlows 37
Big Bang 16–17, 96
Binoculars, types of 28–9
Brahe, Tycho 12

Cassegrain, Guillaume 32
Cassini, Giovanni 14
Catadioptric telescopes ('cats') 34
CCDs (charge coupled devices) 48
Celestial sphere 40
Ceres 79
Chromatic aberration 30
Collimation 33
Comets 76–7
Constellations 21, 102–185
  Andromeda (Princess Andromeda) 110
  Antlia (the Air Pump) 166
  Apus (the Bird of Paradise) 182
  Aquarius (the Water Carrier) 154
  Aquila (the Eagle) 150
  Ara (the Altar) 174
  Aries (the Ram) 134
  Auriga 105
  Auriga (the Charioteer) 114
  Boötes (the Herdsman) 122
  Caelum (the Chisel) 162
  Camelopardalis (the Giraffe) 114
  Cancer (the Crab) 142
  Canes Venatici (the Hunting Dogs) 105, 122–3
  Canis Major (the Great Dog) 162
  Canis Minor (the Little Dog) 138
  Capricornus (the Sea Goat) 154–5
  Carina (the Keel) 166–7
  Cassiopeia (Queen of Ethiopia) 111

  Centaurus (the Centaur) 170
  Cepheus (King of Ethiopia) 107
  Cetus (the Whale) 134–5
  Chamaeleon (the Chamaeleon) 182
  Circinus (the Pair of Compasses) 171
  Columba (the Dove) 163
  Coma Berenices (Berenice's Hair) 146
  Corona Australis (the Southern Crown) 174
  Corona Borealis (the Northern Crown) 123
  Corvus (the Crow) 146–7
  Crater (the Cup) 142
  Crux (the Southern Cross) 167
  Cygnus (the Swan) 130
  Delphinus (the Dolphin) 155
  Dorado (the Goldfish) 163
  Draco (the Dragon) 126
  Equuleus (the Little Horse) 155
  Eridanus (the River) 158
  Fornax (the Furnace) 158
  Gemini (the Twins) 114–5
  Grus (the Crane) 178
  Hercules (hero of Greek mythology) 126
  Horologium (the Pendulum Clock) 158–9
  Hydra (the Water Snake) 142–3
  Hydrus (the Lesser Water Snake) 182
  Indus (the Native American Indian) 178
  Lacerta (the Lizard) 131
  Leo (the Lion) 143
  Leo Minor (the Little Lion) 118
  Lepus (the Hare) 138
  Libra (the Scales) 147
  Lupus (the Wolf) 171
  Lynx 118
  Lyra (the Lyre) 127
  Mensa (the Table Mountain) 182–3
  Microscopium (the Microscope) 178
  Monoceros (the Unicorn) 138–9
  Musca (the Fly) 183
  Norma (the Level) 171
  Octans (the Octant) 183

  Ophiuchus (the Serpent Bearer) 150
  Orion (the Hunter) 139
  Pavo (the Peacock) 174–5
  Pegasus (the Winged Horse) 131
  Perseus (hero of Greek mythology) 105, 115
  Phoenix (the Firebird) 178
  Pictor (the Easel) 163
  Pisces (the Fishes) 135
  Piscis Austrinus (the Southern Fishes) 178–9
  Pyxis (the Compass) 105
  Reticulum (the Net) 159
  Sagitta (the Arrow) 150–1
  Sagittarius (the Archer) 175
  Scorpius (the Scorpion) 175
  Sculptor (the Sculptor) 159
  Scutum (the Shield) 151
  Serpens Caput (the Serpent's Head) 147
  Serpens Cauda (the Serpent's Tail) 151
  Taurus (the Bull) 139
  Telescopium (the Telescope) 179
  Triangulum (the Triangle) 111
  Tucana (the Toucan) 183
  Ursa Major (the Great Bear) 118–9
  Ursa Minor (the Little Bear) 106
  Virgo (the Virgin) 147
  Vulpecula (the Fox) 151
Copernicus, Nicolaus 12

Dark adaptation 26
Declination (Dec) 41
Deep space 80–100

Earth, the 18
Ecliptic, the 19
Einstein, Albert 16–17
Eyepieces 36–7

Fireball (bright meteor) 47

Galaxies 96–9
  'Deep Sky' objects 98
  Andromeda Galaxy, the 98–9
  Black Eye Galaxy 99
  Charles Messier 98
  Coma-Virgo Cluster 99
  elliptical galaxies 97
  irregular galaxies 97
  local group, the 98–9